ENERGY CONSERVATION

Edited by **Azni Zain Ahmed**

INTECHOPEN.COM

Energy Conservation
http://dx.doi.org/10.5772/2788
Edited by Azni Zain Ahmed

Contributors

Mikhail Ja. Ivanov, Dujuan Kang, Vivien Mweene Chabalengula, Frackson Mumba, Mohammed Taih Gatte, Rasim Azeez Kadhim, Akubue Jideofor Anselm, Chong-Hu Wu, A.S. Abdulkareem, A. Jimoh, A.S. Afolabi, J.O. Odigure, D. Patience, U.C. Odili, Lin Rulong, Cai Wenxuan, Xing Bingpeng, Ke Xiurong

Published by InTech

Janeza Trdine 9, 51000 Rijeka, Croatia

Edition 2014

Publishing Process Manager Daria Nahtigal
Technical Editor InTech DTP team
Cover InTech Design team

Additional hard copies can be obtained from orders@intechopen.com

Energy Conservation, Edited by Azni Zain Ahmed
p. cm.
ISBN 978-953-51-0829-0

INTECH

open science | open minds

Contents

Preface

The issue of energy conservation has been with us since the seventies and there has not been a dearth of ideas and increase of knowledge arising from research from various disciplines ever since. The importance of energy conservation has been on the increase that has resulted in the need to produce books to introduce new and richer findings and applications in this field.

This book comprises of nine chapters which cover several disciplines. Each chapter has been carefully written by leading teams of experts from various parts of the world. This publication of book has been intended to look deeply into the rudiments of sources of energy, how energy is transformed from one state into another, from space to ocean. One chapter describes in vivid detail about space energy, cosmic radiation and their relation to established thermodynamic laws. The next chapter describes how energy is collected, stored in the ocean and describes the mechanisms on the energy transformations and how energy can be used from tidal waves. Another chapter actually dwells on the importance of education and how the subject should be approached to make students understand energy and energy conservation better.

The section on applications of energy conservation is very interesting as the contents are rarely published. A report on the investigations into traditional earth sheltered dwellings such as sunken earth houses suggests that there are potentials in passive building insulation which utilizes ground thermal inertia. The techniques of earth sheltering have not yet become common knowledge in building practice and architecture, and as such society is unaware of the process or benefits of this type of building construction. On the contrary, it is well-known that dams are sources of renewable energy but it is not commonly reported about energy losses that can occur in the process of transforming kinetic energy to electrical energy. A chapter on hydropower explains the fundamentals of energy production from dams and how energy can be conserved through the considerations and design of the turbines and pipelines where most of the energy is lost from friction.

Although energy conservation plays a vital role in protecting the environment and optimising the use of energy from conventional fuels, there is of course the contribution from alternative energy resources. A chapter reports on the production biodiesel form groundnut oil as an alternative to petrol diesel. We also cannot deny that industrial processes consume the largest percentage of energy in the world. One chapter provides suggestions on how energy can be conserved in the production of ultra-fine spherical cobalt powders and another on the optimal extraction of microalgal lipid. These are certainly innovative ways to contribute towards energy conservation.

This book does not necessarily cover all the disciplines or the processes that lead to the reduction of energy consumption in the world. However, the ideas and possibilities that they represent are just a minute effort to conserve the energy resources on this planet.

As the Editor of this book, I have found it a pleasure to read each and every chapter and needless to say it was not easy either. With the enormous responsibilities as a one of the top academic administrators of a university, it was an uphill task all the time to complete my duties. I am most honoured and grateful to InTech for appointing me as the editor of this book. In the process, I too, had learned a lot as through the responsibilities as editor my knowledge understanding in energy conservation was enhanced.

I would especially like to thank Ms Daria Nahtigal, the Publishing Process Manager for pushing me all along to meet the deadlines. Finally, I hope that all the readers would find this book refreshing, motivating and, most of all, that the book would be of benefit to students, researchers, professional and practitioners who wish to make this world a better place.

Azni Zain Ahmed
Head
Centre for Research and Innovation in Sustainable Energy
Universiti Tecknologi MARA
Malaysia

Section 1

Understanding Energy Conservation

Chapter 1

Space Energy

Mikhail Ja. Ivanov

Additional information is available at the end of the chapter

http://dx.doi.org/10.5772/52493

1. Introduction

1.1. Foreword

"Our knowledge of the world is guesses and delirium"
Omar Khayyam

Vacuum energy. The present chapter considers a possible application of classical mechanics (more specifically, methods of continuum lightly moving media) for description of the enigmatic Cosmic Energy (CE). We shall not touch energy processes in the Universe conditioned by baryon substance conversions when forming and evolving stellar structures and concentrate on a specific question of modeling dominating CE vacuum that fills free cosmic space everywhere and even in case of absence of our traditional substance. Herewith our approach for CE description differs in principal from the virtual energy concept resting on virtual particles or negative pressure. We consider the CE as a real energy of movement and interaction of vacuum-filling mass particles, which in divers' time were referred to as ether particles or photons (with finite rest mass) or today particles of hidden Dark Matter (DM) [1-6].

Dark energy. Here we specially pay attention to widely discussed nowadays a notion of cosmic Dark Energy (DE) [6-8]. The term "dark energy" appeared in scientific literature in the end of XX Century and marked the cosmic media filling the whole Universe. DE is inseparably linked with any space cubic centimeter and according to well-known formula $E = mc^2$ can be considered a DM equivalent. The first word in terms "dark energy" and "dark matter" means that this matter form allegedly does not emit and does not absorb electromagnetic radiation and interacts with usual matter by only gravitationally. The word "energy" opposes the given media to the structured one consisting of substance particles. DE density, unlike usual and dark, substance is similar in any space point and its pressure has negative value. The negative pressure value is the result of the thermodynamic correlation

ΔE=-$p\Delta V$, which shows the increasing of energy ΔE accompanies by the increasing of volume ΔV. The DE state equation is written as $p / \rho c^2 \approx -1$ [8]. The standard cosmology model gives DE up to 74% of the total mass-energy quantity of our Met galaxy. The figure shows the matter distribution in the Universe accepted today in cosmology. The modern science has carried in consideration the DE and entrusted on it responsibility for the registered accelerated Universe expansion. Thus we display in detail appropriate experimental data.

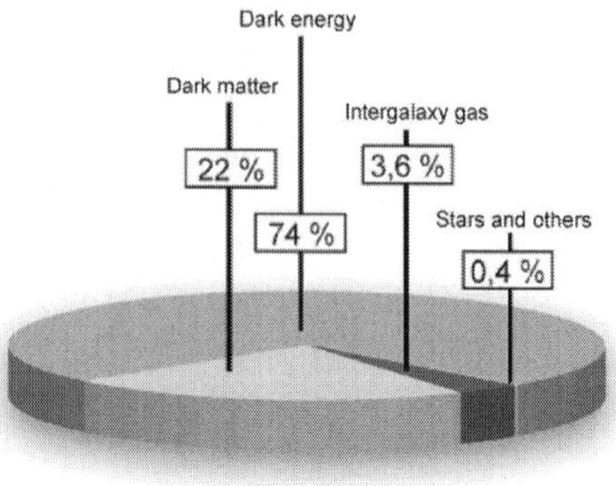

Matter distribution in the Universe

Universe expansion. The last hundred years it is known that Universe expanses. Its possible accelerated expansion was also discussed and particularly popular this subject became since 1998 when the works [9] were published. Observations of distant Super Nova show that galaxies scatter from each other with all greater and greater speed. Today this result is considered without doubts. It was noted as a grandiose achievement of modern physics and awarded a Nobel Prize for 2011 [10].

We think necessary to emphasize that the conclusion about accelerated Universe expansion entirely rests on a postulate about constant light speed in vacuum $c = 2{,}998 \cdot 10^8$ m/s and about impossibility to excess the given speed value by moving bodies. Meanwhile we have to acknowledge that modern science does not dispose reliable data on measured light speed in the early hot Universe. If in hot Universe the light speed in vacuum exceeded its known today value, than the observed accelerated Universe expansion would get a natural trivial explanation. We again return to its usual (not accelerated) expansion and no necessity to carry in consideration the hypothesis of DE presence in Universe. Therefore the given fact of allegedly accelerated should be interpreted as validation of possible light speed dependency on temperature of cosmic vacuum. Note in this connection some published experimental data on measured superluminal light propagation.

Superluminal speed. The history of experimental registration of superluminal light propagation is as old as its prohibition. This phenomenon was thoroughly studied already in 30-s of last Century. Thus in [11-15], in particular, it was stated that in hollow metallic tubes electromagnetic waves can have superluminal propagation. The effects of superluminal propagation of laser electromagnetic pulses were open and studied in 60-s in academician N.G. Basov's laboratory [16,17]. Measured isolated pulses propagation was 6-9-fold light speed in vacuum. In experiment [18] recorded pulse propagation in inverse populated cesium vapor was 310-fold light speed in vacuum. During last 20 years such type experiments [19-21] as well as experiments on superluminal tunneling [22-25] also registered a notable excess of vacuum light speed. Superluminal propagation of centimeter radio waves recorded in [26] and discussed in [27, 28].

Review [29] attempts to explain the specified effects from the standard physical theory. However some published experimental results are rather difficult to interpret in such a way. Some last empirical data on superluminal effects in hot hollow metal tube are presented in [30-33]. The mentioned experimental facts of superluminal propagation of electromagnetic waves and isolated pulses have a principle meaning under building adequate natural physical theories.

Causality. One of the fundamental tenets of modern physics is the sacred causality principle stating that no signal can be transmitted faster than light c. It is confirmed that at speeds higher than c events sequence becomes inverse, the events band allegedly rolls back. However this widely spread misunderstanding itself rest on a postulate that light speed is a maximum possible velocity of interactions propagation. Invalidity of such a statement about violation of the Principle of Causality can be easily shown by the following example. We consider another similar postulate ("sonic postulate") stating that sound speed in a free atmosphere space is a maximum possible velocity of interactions propagation. Then supersonic propagation will naturally violate sonic principle of causality. An unambiguous conclusion: superluminal propagation like supersonic one does not violate the principle of causality.

God particles. Rather full analysis of the question about Cosmic Energy is impossible without regarding links with a fundamental question about the God Particle existence – Higgs boson [34] responsible (as someone thinks) for birth of baryon matter mass. Due to careful last year's experiments at the Tevatrone and Large Hadron Collider this question now is close to its objective decision. "To be or not to be" of the Higgs boson – leading world physics forums solve today [35-37]. Still we have no positive answer for this question, and a possible creation of theoretic physics models are under consideration beyond standard model limits [38] (e.g. models of extra dimension, super symmetry, top quark physics, etc.). However alongside with the mentioned up-and-coming directions one should not forget about classic physics models up to limits of ditto time standardized models. Here the typical example may be the Hidden Mass Boson (HMB) [39, 40], which also as the Higgs boson might take itself the responsibility for the baryon matter birth.

The chapter materials offered to the Reader is entirely based on the traditional classic physics. We show the Cosmic Space contains gaseous medium of HMBs with temperature $T=2.725\ K$ and study in detail this medium. *The Space Energy presents in our case the kinetic energy of HMB particles.* We demonstrate the Grand Unified Theory of electromagnetics, week and strong interactions, electrovalence linkages and antimatter. As additional experimental confirmations we use the Hooke law, the Dulong–Petit law, the thermal expansion law and oth.

1.2. From ancient history

"Namely from Leukippus' schooling the era of atomistic in science has started, which is a theoretical foundation of physics and is continuously developing up to present days. Therefore we can with full right consider Levkippus as the "Father" of theoretical physics"
V. Fistul

Theoretic premises of the present chapter are very close to the old Greek philosopher Leucippus (V century B.C.). In his main work "Great space order" ("Μεγαζ διακομοζ") Leucippus has developed the atomic matter theory for near Earth and Space. His more known student Democritus continued to deeply advance the theory and summarized it on the "Micro space order" ("Μικροζ διακομοζ"). These two original works of materialism allows to speak of the "Leucippus – Democritus line" in the human through development. About a century and a half later old Greek philosopher Epicurus seemed to complete the development of ancient direction of atomic matter theory and publishes a final work "On atoms and vacuum". Their successor poet-philosopher of antic Rome Titus Lucretius Carus (first half in I century B.C.) brilliantly described this theory in his famous poem "On the Nature of Things" ("De Rerum Natura"). Note specially, the developed theory of antic atomism was the most successful theory, actually spread the atomism up to the matter "subbaryonic" level and presented it as an "information bearer" for a human being.

Democritus circa 470 – 390 B.C.

Epicurus circa 341 – 271 B.C.

Lucretius Carus circa 95 – 55 B.C.

In conclusion we cite a brightest inference of ancient materialism by Lucretius Carus [41]:

"In is clear from here that the essence of soul and intellect
Has been created undoubtedly from primary particles petty,
And absence of their gravity does not lessen their essence".

1.3. From XVIII-XIX centuries

"The ether finest substance filling the whole invisible world is capable to possess this movement and heat and also it transfers this movement obtained from the Sun to our Earth and other planets and heat them, so the ether is a medium by means of which the bodies separated from one another transfer heat to each other".
M.V. Lomonosov [42]

The question of space energy and heat nature was thoroughly studied by Russian scientist M.V. Lomonosov in the middle of XVIII Century. In his original work "About the reason of heat and cold" [42] he argues that "the heat consists of substance movement and this movement though not always sensitive but really exists in warm bodies. This movement is internal, i.e. insensitive particles are moving in warm and hot bodies, and bodies themselves comprise these particles." Considering the heat (energy) nature in the form of finest substance particles and ether movement, Lomonosov in ditto time sharply raises an objection to another interpretation of heat nature in the form of hypothetical heat generation (phlogiston), when the body heat is tied with heat generation quantity and heat transportation is connected with heat generation flow from hot bodies to the cold ones. Lomonosov says: "We confirm that the heat cannot be prefixed to concentration of some fine substance no matter its name...". The present chapter demonstrates from modern scientific positions the fairness of many Lomonosov's conclusions regarding the heat (energy) nature.

We want to underline that the heat generation concept (phlogiston concept) in spite of justified critiques has continued to dominate in science for about 100 years. Thus Saudi Carnot leaning on the heat generation model in his wonderful work "Cogitation about fire propulsive force and about machines capable to develop this force" published in Paris, 1824 comes to few basic results of heat theory and formulates a principal idea of heat and work equivalence. He writes in his later researches: "The heat is not that other but propulsive force or rather movement that changed its shape; it is body particles movement; everywhere propulsive force disappears at once the heat generates in qualities exactly proportional to the disappeared propulsive force quantity. Otherwise, always when heat disappears the propulsive force originates." To present the "propulsive force" notion later the scientists started to use the term "energy". We give also Carnot's formulation of the energy conservation law: "Thus one can say a common expression: the propulsive force exists in the nature in constant quantity; generally speaking, it is never produced, is never destroyed; actually it can change its shape, i.e. it generates either one movement type or another, but never disappears". We find further development of the energy conservation thermodynamic law in works by Mayer, Joule, and Helmholtz the latter gives its generally accepted mathematical formula.

M. V. Lomonosov (1711-1765)

Saudi Carnot (1796-1832)

Rudolf Clausius (1822-1888)

Beginning with R. Clausius and W. Thomson studies in 50-s of XIX Century the heat nature is finally connected with certain particles movement type. The basic Clausius work "Mechanical heat theory" considers "light heat as ether fluctuating movement"; in bodies "the heat is movement of substance and ether fine particles", and "heat quantity is a criterion of this movement live force". R. Clausius developed the basics of classic thermodynamics and proposed a convenient form of their presentation as the first and second beginnings.

W. Thomson in his studies on heat theory and, in particular, in work "About heat dynamic theory" also believes that "heat is not a substance but a dynamic form of mechanical effect" and "certain equivalence should exist between work and heat". In contrast with R. Clausius, W. Thomson especially pinpoints an idea that the second thermodynamic beginning expresses the energy dissipation process. Thus, W. Thomson believes, the nature is under control of the "energy dissipation principle". The idea of "the Universe heat death" is known follow from this principle.

The second half of XIX Century is marked by outstanding James Clerk Maxwell and Ludwig Boltzmann works on the theory of thermogasdynamic processes. In review monograph "Substance and movement" Maxwell gives the following definitions: "The energy is an ability to conduct work", "The heat is an energy type" and explains that "warm body fine particles are in a state of quick chaos agitation, i.e. any particle always moves very quickly and its movement direction changes so fasten that it displaces very small or does not displace at all from one place to another. This is true, a part, and may be very big part of warm body energy should be as a kinetic energy type".

In addition to the mentioned citing it is very important for further discussion the Maxwell's weighted velocity wise distribution of similar particles system uniquely determined by system temperature and particles mass. Presently the formula has a title: the Maxwell-Boltzmann equilibrium velocity distribution formula. Our work main conclusions in much degree are based on this distribution.

William Thomson (1824-1906)

James Clark Maxwell (1831-1879)

Ludwig Boltzmann (1844-1906)

Principal questions of thermodynamic theory from standpoint of gas kinetic theory are studied by Ludwig Boltzmann. His "Lectures on gas theory" [43] begin with words "Already Clausius strongly distinguished the general mechanical heat theory based mainly on the two theorems, by his example called the elements of heat theory, from the special theory, which, firstly, certainly believes that heat is a molecular movement and, secondly, attempts to develop a more accurate interpretation for this movement character." Using statistic approach Boltzmann analyzes main positions of the mechanical heat theory.

Completing the review of energy theory origins we additionally underline that foundations of electromagnetism theory are obtained by Maxwell also on the base of ether concept [44]. It is well known, Maxwell in his basic "Treaties on electricity and magnetism" used a model of light ether as an invisible fluid. In particular, in the article "About Faraday power lines" he writes: "Reducing everything to purely mechanical idea of some imaginable fluid movement I hope to achieve generalization and accuracy and avoid the dangers that occur at attempts with the help of premature theory to explain phenomena reasons."

Discussion on questions of physics basics, and first all, "what actually exists - matter or energy", unrolled sharply on the edge of XIX and XX centuries. In a bright report "On development of theoretic physics technique in the newest time" delivered at a Nature Scientists Meeting in Munich, September 22nd, 1899 Ludwig Boltzmann considers this question from the positions of classic atomistic mechanics. Touching electromagnetic phenomena he, in particular, says: "The matter is that alongside with weighted atoms the existence of special substance made of significantly smaller atoms, namely, light ether been allowed. In turned out possible to explain nearly all right phenomena, which earlier Newton referred to do a specific light particles emanation, by lateral ether oscillations. Though some difficulties remained, for instance, it was unclear why there is complete absence of longitudinal waves in the light ether, which in all weight bodies both exist and play major role". Further in that report, expressing care about the approaches of newest "energetic" physics being published Boltzmann exclaims: "I have remained the last among those who admitted the old by all their soul; else I am the last who still struggles for this old whenever possible. I see my vital task as the following: by means of possibly clear and logically regulated development of the old classic theory results to make if in future not-necessary to re-discover a lot of good and still suitable things, that, in my opinion, are contained in this theory, as it repeatedly happened in this history of science". Let's cite as well another Boltzmann's idea said in the end of that report: "Whether once again would the mechanistic philosophy with the battle, having at last found a simple mechanic model of light ether, at least would the mechanic models maintain their meanings in future or would a new non-mechanic model be accepted the best?"

1.4. From XX century

Sir, I tell you on the level:
We have strayed, we've lost the trail.
What can WE do...

A.S. Pushkin (1830)

The mechanical philosophy has suffered a major defeat from creation of special and general theories of relativity. The authors of basics of the special theory of relativity (STR) rested entirely on the two widely discussed scientific results of that time. The first followed from unsuccessful experiments on detection of light-bearing ether (famous experiments by Michelson – Morley). A. Poincare, one of the STR authors writes about it in 1895: "Experience has given a lot of facts, which admit the following generalization: it is impossible to detect absolute matter motion or, more exactly, relative motion of material matter and ether. All that is possible to do – to detect material matter motion relative to material matter". The principal generalization cited becomes further the first STR postulate. Below this postulate is given from P. Tolman monograph [45] approved by A. Einstein himself: "It is impossible to detect the uniform progressive motion of a system in free space or in any hypothetical ether media, which could fill this space". Therefore it is reasonable to speak only about relative movement of two systems and no sense to mention absolute motion (for example, absolute system motion relative to the universe ether). From this viewpoint, there is no necessity in the mechanic ether.

The second scientific result deals with the field of theoretical research of that revolutionary time. For the first time these linear transformations of space-time coordinates named after Lorentz were obtained in 1887 by V. Fogth as transformations preserving the invariance of d'Alembert wave equation. In 1900 G. Larmour showed that the Maxwell's electrodynamics equations in free space are also invariant regarding these transformations. Later in 1904 they also were written by Lorentz and since 1906 by Poincare proposal were named the Lorentz transformations. Considering peculiarities of the Lorentz transformations Einstein points out: "Velocities of material bodies exceeding speed of light are impossible, that follows from appearance of the radical $\sqrt{1-v^2}$ in formulas of the particular Lorentz transformation". Assuming, that at transition to a system moving with v speed the time-space coordinates changes in accordance with the Lorentz transformations, quite naturally the second STR postulate becomes effective: "the speed of light in free space is identical for all observers regardless relative velocities of a light source and observer". The speed of light c in free space may now be considered as the maximum allowable speed and, first of all, the maximum speed of interactions propagation. "Unification of the relativity principle and limit speed of interaction propagation is called the principle of relativity by Einstein (it was formulated by Einstein in 1905) is contrast with the principle of relativity by Galileo, which is based upon the infinite velocity of interaction propagation,"- read the "Field theory" course L.D. Landau and E.M. Lifshitc [46].

Thus the STR erects an insurmountable "light barrier" for allowable matter motion speeds and for any weak or strong signals propagation speeds (speeds of information propagation). In completely rejects the body ether of XIX century. The Einstein's article "About the ether" begins with the words: "Speaking here about the ether we surely do not mention the physical ether of mechanic wave theory which obeys the Newton laws and some points of which have velocity. This theoretic presentation, in my opinion, has completely gone off the scene with the creation of the STR". Roughly so reads a verdict to the classic physical ether.

The ideas of STR are developed in the General Theory of Relativity (GTR) created by Einstein in 1906 – 1915. The physical processes in GTR are described in a curved space-time system with variable metrics. The curvature of space-time allows to build an interesting model of gravitational interaction based on similarity between the results obtained in gravitational field using a uniformly moving reference system and results in absence of a gravitational flied with an accelerating reference system. This similarity is accepted as the second GTR postulate and called the equivalence principle. The GTR realizes in full the idea of relativity of any motion.

Further physics development in the XX century was dominated unconditionally by the GTR and STR and, naturally, was brightly imposed on the whole modern theoretical physics.

2. Key experimental data

"A Rational Being! Turn your eyes to the serene sky.
What a wonderful order is there!"
Kozma Prutkov (Russian man of sense, literary person).

2.1. Cosmic microwave background radiation

One of the important achievements of experimental astrophysics of the second half of the XX century was the discovery of the Cosmic Microwave Background (CMB) radiation. The existence of this radiation was predicted by George Gamow in 1948. According to his idea of the Universe origin as a result of the Big Bang, the current radiation appeared at the initial stages of the Universe development and it was "severed" from the matter spreading this radiation and so far it has cooled off to a very low temperature. The temperature of CMB was predicted by Gamov also (accurate within 7 K). In 1955 the post graduent radio astronomer T.A. Shmaonov in the Pulkovo Observatory experimentally observed noise microwave radiation with the absolute value of the effective temperature 4 ± 3 K. After Shmaonov had defended his dissertation, he also published the article [47].

In 1966 CMB was registered by the American astronomers A. Penzias and R. Wilson [48]. The careful researches of the last decades showed that the distribution of the radiated density does not depend on the direction of its registrations, and that it corresponds to the equilibrium radiation of a black body with the T=2.725 K. These properties of radiation discuss the possibility that we don't have the case with transformed radiation of stellar objects but with the independent substation filling the entire Universe. According to Ja.B. Zeldovich [49] CMB was sometimes called "new ether". This name was given as a result of dipole anisotropy discovery of CMB at the middle of 70-years [50]. This circumstance allows the introduction of the absolute cosmological frame of reference in the vicinity of our own Galaxy where the background radiation is isotropic (accurate within small-scale fluctuations).

Before the discovery of CMB with the final value of the temperature T it was thought that the temperature in the vacuum of cosmic space is $T=0$ K and the pressure of the vacuum is $p=0$. These values conformed to the properties of carries of electromagnetic radiations – photons with their rest mass $m=0$, the velocity of their moving in a free space can be equal to the speed of light in vacuum $c=2.998\cdot10^8$ m/s only, their impulse P and energy E that connect with each other by the formula $E=Pc$. The photons do not cooperate with each other, and their totality behaves as ideal gas (with the adiabatic index $\kappa=4/3$).

However, the discovery of the final temperature of CMB T=2.725 K should cardinally change the situation. By virtue of the kinetic theory by L. Boltzmann [41] and the dimensional analysis [51-53] (the π - theorem by E. Buckingham [54]) we come to the final values of pressure and the mass particle in the physics space vacuum

$$p = nkT, m \sim \frac{kT}{c^2},$$

where k – the Boltzmann's constant, n –concentration of particles in examining space. Thereby we introduce with necessarily the ideal gas mass medium to the cosmic space (physical vacuum). This medium can be identified with the mass photon gas or the Dark Matter (DM) of the 20th century or it can also be considered as the classic ether of the 19th century. In the present work for such a medium the particle structure is based and called Hidden Mass Boson (HMB).

2.2. Dark matter

We shall present the main experimental facts proving existence of Dark Matter (DM), which were repeatedly described in literature, (see [1-7]).

Astronomer Oort set the problem of possible presence of DM in the Universe in 1932 with connection to his measurements of motion of stars in the disk of our Galaxy Milky Way. An unexpected conclusion of his measurements was an essential lack of the aggregate Galaxy weight for explanation of rotational speeds of the disk. One year later astronomer Zwicky, studying dynamics of the clusters of galaxies, has come to the conclusion, that the observed weight makes only about 10% of the total mass required for a reasonable explanation of observations of the total mass in the clusters of galaxies. Special spectroscopic and radioscopic observations of rotational speed of hundreds of spiral galaxies were executed later. These observations have shown the essential increase of total mass of galaxies in the direction of the edge of the stars disk. These facts indicated to presence of the spherical material halos surrounding the spiral galaxies, which could not be registered in the other way (i.e. was invisible). Careful observations of elliptic galaxies and clusters of galaxies have also indicated to the presence of the invisible dark matter gravitationally interacting with the visible objects of the Universe.

The other experimental confirmation is the hot gaseous congestions. Explanation of their existence with registered parameters requires presence of DM in quantities noticeably exceeding the visible matter. We underline, that the empirical fact of the DM existence in Meta-Galaxy is now conventional.

DM at present time is one of the most intriguing mysteries of the nature. There were multiple attempts to describe the nature of DM, but no one was successful yet (see [4, 5]). As it was already noticed, DM (with DE) makes not less than 96% of all matter of our Meta-Galaxy. A number of theoretical models of DM based on principles of the modern physics are proposed. Spreading of mass value of the DM particles in different theories reaches 78 to 80 orders: from mass values of 10^{-6} eV for ultra-light (sterile) axions up to values of about 10^{6} M_0 (M_0 is the mass of Sun) for supermassive black holes (in kilograms this range corresponds to mass values from 10^{-42} to 10^{36} Kg).

We shall briefly list the main candidates for the DM particles. The main candidates for the baryonic DM are the Massive Astronomical Compact Halo Objects – MACHO. Usually MACHO includes dwarf stars (white, black or brown dwarfs), planet such as Jupiter, neutron stars and black holes. The careful analysis of the aggregate contribution of the listed baryonic objects to the total substance mass of the Universe performed last years has shown that the DM problem cannot be solved with help of the baryonic matter only. We repeat that the aggregate estimation of the all baryonic matter in the Universe makes less than 5% of the total matter.

Non-baryonic DM is usually divided into two main categories: Cold Dark Matter – CDM, which particles moves at subluminal velocities, and Hot Dark Matter – HDM, which particles should move at relativistic velocities, close to that of light in vacuum. The main candidates for CDM should be Weakly Interacting Massive Particles – WIMP with mass of 1 GeV and more. Now intensive searches of such particles with the help of different special detectors are under way [5-7].

The ultra-light sterile axions with mass in range of $10^{-6} \div 10^{-3}$ eV are considered as hypothetical DM particles. Axions were introduced into the theory of elementary particles in order to explain violation of the CP–invariance in the early Universe [55, 56]. Axions should breakdown into two identical photons under action of an external electromagnetic field [56]. With the purpose of registration of such disintegration and confirmation of the axions existence, the super-sensitive axion detectors are developed now.

Neutrino and anti-neutrino are the characteristic particles of HDM. Their mass, as candidates for HDM, should be in the range of 10 to 100 eV.

Other candidates for DM are different super-symmetrical particles, which could be formed at the initial moments of the Universe existence (when it was "super-symmetrical" and when all the four interactions were united). Detailed information on all possible hypothetical DM particles is presented also in the proceedings of the last conferences on particle physics [35-38].

2.3. Some astrophysics data

Gamma-ray bursts. An actual unsolved problem of astrophysics is the problem of powerful bursts of gamma radiation, which registration began in the middle of 60's with the help of special equipment of reconnaissance satellites [57, 58]. Now similar bursts are registered 3 to 5 times a day and are intensively studied [59-62]. An important feature of the gamma–bursts is their after-glowing in x-ray and light ranges, which approaches Earth considerably later than initial bursts do.

The delay time of after–glowing can be more than one year [60, 61]. At the present time there is no theoretical model, adequate to the phenomena of the gamma–bursts propagation to the large cosmological distances, which could explain absence of apparently essential energy losses of the propagating bursts.

Cosmic rays. Origin of the cosmic rays of ultra–high energy (more than 10^{20} eV) is a difficult to explain problem of astrophysics [63-65]. The registered energy level greatly exceeds the permissible by theory limit of the energy spectrum of particles of the primary cosmic rays (because of the known effect of interaction with relict photons of Greisen–Zatsepin–Kuzmin (GZK) [66]). Unanswered questions are both the mechanism of the charged particles acceleration to the high energy levels and capability of their propagation to the huge cosmological distances without essential energy losses. The explained problem is called "GZK paradox". Theoretical approaches beyond the framework of the modern standard models of theoretical physics are proposed (see, for example [35]).

Cosmic jets. The extragalactic "superluminal" jets propagating from the centers of quasars and active galaxies remain an astrophysical enigma [67]. Registered during the last years velocities of propagation of such jets exceed the speed of light in vacuum by 6 to10 times [68–70]. One of the latest publications on this subject informs of the presence of a giant jet of more than 300 kilo parsec of length (one million light years) at quasar RKS 1127–145 containing more than 300 "barrels" [71].

In the last time during two decades, there is considered also the mentioned earlier accelerating expansion of the Universe. The accelerating expansion means that the Universe could expand forever until, in the distant future, it is cold and dark. The team's discovery led to speculation that there is a "dark energy" that is pushing the Universe apart [9, 10]. Astronomers and physicists have so far failed to discover the nature of this strange, repulsive force with negative pressure [8].

2.4. Electromagnetic evidences

Sonoluminescence. Sonoluminescence is the emission of short bursts of light from imploding bubbles in a liquid when excited by sound [72, 73]. Sonoluminescence can occur when a sound wave of sufficient intensity induces a gaseous cavity within a liquid to collapse quickly. This cavity may take the form of a pre-existing bubble, or may be generated through a process known as cavitation. Single bubble sonoluminescence occurs when an acoustically trapped and periodically driven gas bubble collapses so strongly that the energy focusing at collapse leads to light emission (Figure 1). Detailed experiments have demonstrated the unique properties of this system: the spectrum of the emitted light tends to peak in the ultraviolet and depends strongly on the type of gas dissolved in the liquid; small amounts of trace noble gases or other impurities can dramatically change the amount of light emission, which is also affected by small changes in other operating parameters (mainly forcing pressure, dissolved gas concentration, and liquid temperature). The light flashes from the bubbles are extremely short—between 35 and a few hundred picoseconds long—with peak intensities of the order of 1–10 mW. The bubbles are very small when they emit the light. Transfer of shock wave kinetic energy to light burst energy may be simulated with help of the conservation law system from the section 4.4 of the chapter.

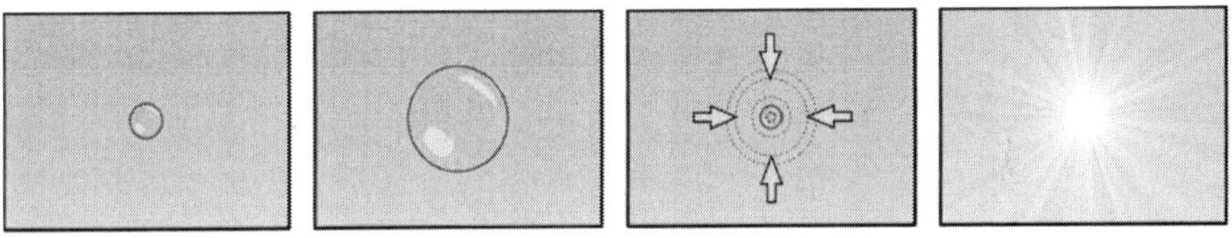

Figure 1. Sonoluminescence main stages

Displacement current. The proposed in this chapter physical model of vacuum and the suggested structure of the HMB give very clear interpretation of the Maxwell displacement current and the Umov-Poynting vector. Firstly, we are reminded of the spread example of necessity to introduce the displacement current to Ampere's law. For instance, a parallel plate capacitor with circular plates is charged by current I (Figure 2). The magnetic field in the point at a distance r from the conductor can be calculated by Ampere' law

$$\oint \bar{B} d\bar{s} = 4\pi \frac{k_0}{c^2} \int_S \bar{j} d\bar{A}.$$

Integrating with respect to the circle we will obtain the magnetic field in the point P

$$\bar{B} = \frac{2kI}{c^2 r},$$

where $I = \int_S j dA$ is summed current flowing through the surface S. The Ampere's law has to issue for surface S', that is based on the same circle but coming between plates of the capacitor. By adopting this method the availability of current and electric field between the plates of the capacitor, the displacement current is introduced to Ampere's law has an additional term

$$\oint \bar{B} d\bar{s} = 4\pi \frac{k_0}{c^2} \int \bar{j} d\bar{A} + \frac{1}{c^2} \int \frac{\partial \bar{E}}{\partial t} d\bar{A}$$

The last term was introduced by Maxwell (and was named the displacement current). The physical meaning of the displacement current in our case is the displacement of dipoles of HMB and their re-orientation between the plates of the capacitor. In vacuum (when dielectric is absent) availability of the displacement current between the plates of the capacitor comes to the polarization of real HMB dipoles and we see clean and real physical interpretation of this current.

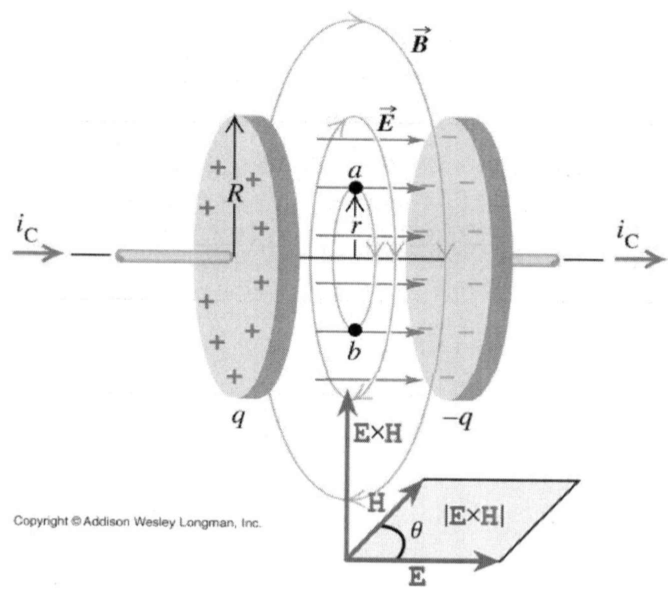

Figure 2. The Maxwell displacement current and the Umov-Poynting vector.

The Umov-Poynting vector. Now we examine the Umov-Poyting vector of the electromagnetic energy flux. With changing of electric field E within capacitor, its internal volume acquires energy. The energy flux is described by the Umov-Poynting vector $\overline{E} \times \overline{B}$ and has direction from the edges of the plates to inside of the capacitor (Figure 2) from its surrounding space. As well as the displacement current in our case (availability of the HMB dipoles) the energy flux of electric field has physical meaning and connects with the moving (flowing) of dipoles from surrounding to inside.

The propagation of current inside the conductor takes place when the incidence of electric potential along the conductor and the availability drops electric field $\overline{E}$ inside and close to the conductor which in turn is directed alongside the conductor. Availability of the current induced magnetic field $\overline{B}$ around the conductor is directed on tangent to the circle around the wire. The vectors $\overline{E}$ and $\overline{B}$ are perpendicular. There is energy that flows to the inside of the conductor from all sides and which accompanies an energy lost by the conductor as heat.

The classic theory shows that electrons take their energy from outside, which means from the energy flux of the outside field to the inside of the conductor. Existed HMBs in electric dipole form are created by the energy flux going to the conductor, which in turn is processing electric current along it more clearly. The same natural interpretations we get to known phenomena of inductance and self-inductance.

3. Key theoretical items

"And for the rest, summon to judgments true,
Unburied ears and singleness of mind
Withdrawn from cares; lest these my gifts, arranged
For thee with eager service, thou disdain
Before thou comprehendest: since for thee
I prove the supreme law of Gods and sky,
And the primordial germs of things unfold,
Whence Nature all creates, and multiplies
And fosters all, and whither she resolves
Each in the end when each is overthrown."
Titus Lucretius Carus [41]

Theoretical models of the nonlinear dispersion processes considered below are wholly based on the experimental facts and principles of the Newtonian mechanics. When describing these processes, the independent variables are time t and three spatial coordinates x, y, z, i.e. motion of a material medium is supposed in the three-dimensional Euclidean space. Taking into account fundamentality of concept of time and space and broad discussions of this problem held in the XX century, we shall briefly consider this and other problems which are important from the point of view of the present chapter (in particular, concepts of the subbaryonic dark matter and dark energy, principle of relativity and some other principles).

3.1. Time

"Absolute, true, mathematical time itself and its essence,
without any relation to anything external flows
uniformly and in another way called duration"
I. Newton "The Beginnings" [74]

"The time does not exist by itself, but within the subjects,
We all feel it, when something happened in the past,
Whether it happens now, or in the future it will follow…
So far, one has never considered the time
Without its relation to the bodies' motion or their sweet rest."
Lucretius Carus "De Rerum Natura"[41]

Absoluteness and one-dimensionality. In the Newtonian mechanics, the basic property of time is its absoluteness. This property means that time does not depend on the selected reference system. Any time interval Δt is the absolute invariant for any nature process and does not depend on course of the considered processes. It is convenient in many cases to accept a second as a reference time unit and to define it as a basic unit for the time measurement.

Time is one-dimensional and common always and everywhere.

Homogeneity. Another important property of time is its homogeneity. Homogeneity of time characterizes its uniform development and supposes equivalence of all moments of time. This property allows to arbitrary select the zero time reference, i.e. an instant t=0, and reference interval of the time measurement. It is convenient to study the development of any process from its initial reference point and to characterize it by the current moment t, measured by the selected interval (measurement unit). It is also convenient to connect the absolute time with our Universe and assign the initial reference point to the moment of the Big Bang.

Single directivity. Property of single directivity of time consists in its development in one direction only – forward, i.e. increasing its value. It cannot be turned back to the past. In other words, there is a so-called arrow of time [75] (from the past to the future). Thus the principle of causality is fulfilled – corollary depends on the reasons, causing that, i.e. the present order of things depends on the past only and does not depend on the future. We underline, that in our case the causality principle is fulfilled unconditionally and irrelevant to any reference value, for example, to the speed of light in vacuum (velocity of propagation of a signal or disturbances). Within the framework of the Newtonian mechanics, the speed of light in vacuum is not the maximal possible velocity of the information transfer.

Abstractness and infinity. Property of abstractness of time is perfectly formulated by Lucretius Carus in the epigraph quoted at the beginning of this paragraph. The property of infinity speaks about absence of the beginning and end of the time flow.

So, we shall use the absolute, one-dimensional, homogeneous, unidirectional, infinite, abstract definition of time. Such concept of time lies in the foundation of the Newtonian mechanics. It was widely used at the dawn of the scientific development in the natural philosophy (Leucipus, Democritus, Epicurus, Lucretius Carus). Transition from the "mathematical" abstract time to the "physical" one is related to selection of the reference point of counting and interval of time measurement.

3.2. Space

"Absolute space by its essence,
Regardless to anything external,
Remains always identical and fixed..."
I. Newton "The Beginnings" [74]

"There is no end of any side of the Universe,
For otherwise it surely would have the edges;
Nothing can have the edges, obviously, anything, if only
Outside of it there is nothing, that it separates,
That would be visible
Our feeling was, up to monitor it is capable...
Also is indifferent, in what you are parts installed..."
Lucretius Carus "De Rerum Natura"[41]

Absoluteness, three-dimensionality, isotropy and homogeneity. Rather customary for our usual life the absolute three-dimensional Euclidean space forms the basis of the Newtonian classical mechanics. It is isotropic and homogeneous. All points of the space and all directions are equivalent. These properties of space allow to arbitrary select a point of origin of three independent Euclidean coordinates $x,y,z,$ (for example, rectangular Cartesian) and direction of the coordinate axes. It is convenient to combine the coordinate origin of the absolute space with an initial point in space, where in limit Big Bang was concentrated. Thus, the problem of selection of direction for the coordinate axes of isotropic absolute space will not be essential. All other coordinate systems should be considered relative to the selected unified absolute system.

Abstractness and infinity. Space is abstract and infinite. Similarly to time it "does not exist by itself, but in subjects we feel it". Space is infinite in all directions. Similar to the time variable, the transition from the "mathematical" abstract space to the "physical" one consists in selection of a system of spatial coordinates and unit of length measurement.

Invariant interval. The main geometrical invariant when transiting from the absolute system of Cartesian coordinates to another Cartesian system will be value

$$\Delta s^2 = \Delta x^2 + \Delta y^2 + \Delta z^2.$$

which uniquely determinates distance $|\Delta s|$ between two points in space. Thus transition from some coordinate system to another one is carried out at fixed time, when location of points in space is fixed.

The basic result of brief consideration of concepts of time and space is the following. All natural processes under investigation are characterized by four independent variables: by absolute time t and three absolute Euclidean coordinates x,y,z. When transiting between two coordinate systems, the invariant values are the time interval Δt between two events at fixed point and the spatial interval $|\Delta s|$ between two points in space at a fixed moment of time. Point of origin of the absolute time and origin of the absolute system of spatial coordinates and their directions one can choose arbitrary, basing upon convenience of studying of physical process. As we already noticed, in case of our Meta-Galaxy (visible Universe), it is convenient to consider the absolute coordinate system with the time and spatial coordinates origin located at the initial point of Big Bang.

Experimental data. We shall discuss some experimental confirmations of the Euclidean properties and of three-dimensionality of our space. The last careful experiments on small-scale fluctuations of the relict microwave radiation have shown its isotropy with precision up to 10^{-4} K and have confirmed the Euclidean (plane) geometry of the Universe. Distribution of angular spectrum for power of fluctuations of relict radiation has the significantly expressed first peak, which corresponds to an angle of 1 degree that confirms the plane geometry of our Meta-Galaxy [76, 77].

Three-dimensionality of space is corroborated by validity of the law of reversed squares in the theory of gravitation, electrostatics and hydrodynamics. Spherically symmetrical vector field $\overline{v}$ coursed by a source and depending on radial coordinate R only, is determined by the equation

$$\text{div } \overline{v} = \frac{1}{R^2}\frac{\partial}{\partial R}(R^2 v_R) = 0$$

and has solution $v_R = C / R^2$, where C is a constant dependent on the source intensity.

Area of the closed spherical surface of radius R is equal to $4\pi R^2$ for the three-dimensional case only. The area of the hyper-spherical surface is proportional to other degrees of radius for spaces with other quantity of dimensions. Fulfillment of the Newton's gravitation law, Coulomb's law and law of the velocity field of a source and drain for incompressible fluid in our space (law of reversed squares) proves the property of three-dimensionality of the space.

3.3. Matter

"So, there is nothing where the matter from Universe
Could be disappeared and from where suddenly bursted."
Lucretius Carus "De Rerum Natura"[41]

Materialization of concepts of time and space occurs with help of the real matter filling the space and changing with time. Basing upon the experimental facts, accumulated by the present moment, within the framework of the chapter, we shall also consider subbaryonic material substance along with the usual and customary baryonic matter.

Structure of matter. The percentage composition of the matter in our Meta-Galaxy (observable Universe) is graphically represented in figure 3. Numerous experimental data and a number of theoretical models reliably prove such structure. Metals compose approximately 0.01% of all matter in the Universe. Visible part of the baryonic matter makes about 0.5% only of the total amount of the registered matter. Estimation of all baryonic matter including invisible one (invisible planets such as Jupiter, black holes etc.), gives near 4% of the total amount. The remaining overwhelming part of the matter in the Universe, approximately 96%, is invisible, i.e. cannot be registered in optical range of electromagnetic radiation (as well as in other ranges: x–ray, ultra-violet, infrared, gamma and radio ones). In this respect, this matter in the Universe is called Dark Matter – DM. Approximately 22% of DM concentrated around the galaxies and their congestions in the shape of "spherical atmosphere" (galaxy halo). The rest of DM, about 74% of the total matter, as it is supposed, is rather uniformly distributed in free space of our Universe. Last years this part of DM is identified with Dark Energy – DE or with special Quintessence (see, for example [8]).

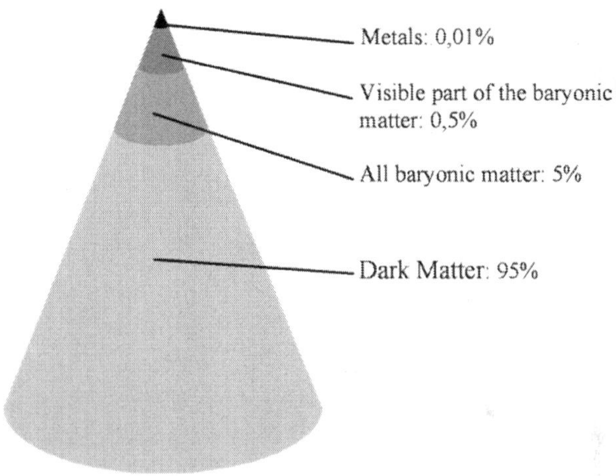

Figure 3. Composition of the matter in the observable Universe.

Baryonic matter. The main elementary components of the baryonic matter are usual protons and neutrons (baryons). Here we merely note that now baryonic matter in the Universe stay mostly (more than 95%) in gaseous (plasma) condition. About 70% of the baryonic matter in the Universe represented by the atomic hydrogen, about 25% is a share of helium. This baryonic matter is concentrated in hot gaseous congestions around galaxies. Theoretical description of dynamics of such hot congestions with good approximation can be conducted with help of methods of gas dynamics and plasma physics. The key moment here is definition of the state equation of the considered gaseous medium, adequate to the nature, especially at rather high temperature and pressure.

Dynamics of ordinary baryonic gaseous medium is described by three conservation laws – mass, momentum and energy. Boundary conditions and the state equation of the gaseous medium close the task. As an example, we shall put here elementary equation of state for the perfect gas

$$p = \rho RT,$$

where p is the pressure, ρ is the density, T is the temperature, R is the universal gas constant. The characteristic speed of the disturbance propagation – speed of sound – is determined in adiabatic medium as

$$c^2 = \frac{dp}{d\rho} = \kappa \frac{p}{\rho} = \kappa \frac{k}{m} T,$$

κ is the adiabatic index, k is the Boltzmann constant ($k=R/N$, N is the Avogadro number), m is the mass of particles of the gaseous medium. Knowing speed of the disturbance propagation and medium temperature one can determine the mass of particles, composing the homogeneous gaseous medium. In the kinetic gas theory the state equation is obtained as a result of application of the Newton's laws.

3.4. Motion

Absolute motion

"The absolute motion is movement of bodies from its absolute location to another one, relative motion - from relative to relative one ... Instead of absolute locations and motions one use relative; in every day's business it does not make an inconvenience, in philosophical one it is necessary to abstract from the feelings."
I. Newton "The Beginnings" [71]

The absolute motion is considered as a movement of a body from one position to another in a selected absolute three-dimensional Euclidean space. As such absolute space we shall accept the recommended in 2.2 absolute Cartesian coordinates system, which point of origin is set to the initial point of Big Bang. The Galilean principle of relativity allows establishing unequivocal conformity between absolute and relative motion.

Axioms of motion

"Axiomata sive Leges Motus.
I. Anybody remains at rest or at uniform and rectilinear motion, unless and until it is forced to change this state by the applied forces.
II. Variation of the momentum is proportional to the applied propulsion force and directed along the straight line of the force direction."
I. Newton "The Beginnings" [71]

Axioms of motion were precisely formulated by Newton and are explained in the introduction courses to the general physics.

Newtonian principle of determinacy. The Newtonian principle of determinacy consists of the following. Combination of positions and velocities of its points set the initial state of a system. It distinctively defines the subsequent motion of the system. In particular, the acceleration is uniquely determined

$$\frac{d\overline{v}}{dt} = \overline{F}(t,\overline{r},\overline{v}).$$

We shall set some restrictions on the right part of equation using for this purpose the space-time properties. Homogeneity of time requires function $\overline{F}$ to be dependent on the relative time $\Delta t = t - t_0$ only, since the transformation of time shift (in other words, selection of the origin point of time t_0) should not change the solution of the equation. Further, the fundamental property of homogeneity of space requires function $\overline{F}$ to be independent on absolute coordinates. Invariance of the solution of the equation regarding the shift in space displays, that function $\overline{F}$ may depend on relative coordinates $\Delta\overline{r} = \overline{r} - \overline{r}_0$ only. Fundamental property of isotropy of space, i.e. the equality of all directions in space, imposes the following requirement on $\overline{F}$

$$\overline{F}(t,G\overline{r},G\overline{v}) = G\overline{F}(t,\overline{r},\overline{v}),$$

where G is the orthogonal transformation of spatial coordinates and velocities. One more requirement to the right part of the equation follows from the properties of invariance with regards to the transition to the uniformly moving coordinate system. The Galilean principle of relativity and the first Newton axiom of motion formulated on its basis works here. Requirement of dependence of function $\overline{F}$ on relative speed $\Delta\overline{v} = \overline{v} - \overline{v}_0$ only follows from this principle. Because of this, the equation should be written in the form

$$\frac{d\overline{v}}{dt} = \overline{f}(\Delta t,\Delta\overline{r},\Delta\overline{v}).$$

$$\overline{f}(\Delta t,G\Delta\overline{r},G\Delta\overline{v}) = G\overline{f}(\Delta t,\Delta\overline{r},\Delta\overline{v}).$$

The elementary examples of these equations are the equations of motion of a body of mass m under constant force $\overline{f}$

$$m\frac{d\overline{v}}{dt} = \overline{f},$$

or equation of motion of the ideal (inviscid) fluid (gas)

$$\rho\frac{d\overline{v}}{dt} = -\text{grad } p,$$

where ρ is the density and p is the medium pressure etc.

Further when building the theoretical models of a nonlinear medium dynamics we shall always satisfy the listed requirements, which follow from the fundamental properties of space-time and the Galilean principle of relativity.

4. Space energy & hidden mass boson

Give me matter, and I will construct a world out of it!
Immanuel Kant (1755)

The Space Energy presents in our case the kinetic energy of HMB particles.

4.1. Dimensional analysis for hidden mass boson medium

We apply an ordinary physical dimensional analysis for free cosmic space in a standard way [51-54]. Executed dimensional analysis bases on measured values of the following quantities: the light speed $c_0 = 2.998 \cdot 10^8$ m / s, the CMB temperature T_0= 2.725 K and the assessment of the critical density of the Universe $\rho_0 \sim 10^{-26}$ kg/m^3. It also uses three well-known constants: the Boltzmann constant $k = 1.38 \cdot 10^{-23}$ kg (m/s)^2 /K, the gravitational constant $G = 6.67 \cdot 10^{-11}$ Nm^2 /kg^2 and the Planck constant $h = 6.63 \cdot 10^{-34}$ J s. Applying π - theorem by Buckingham [54], we construct the dimensionless parameters π and associated physically meaningful relationships.

The simplest dimensionless parameter π_1, connecting pressure, density and velocity, can be written as

$$\pi_1 = \frac{p}{\rho c^2} \sim 1$$

and allows an assessment of the "critical" pressure value

$$p_0 \sim \rho_0 c_0^2 = 10^{-9} \text{Pa}.$$

In the case of an ideal "photon" gas with adiabatic constant $\kappa = 4/3$ the speed of perturbations is defined as

$$c_0^2 = \kappa p_0 / \rho_0$$

and then we get $p_0 = 1.4 \cdot 10^{-9}$ Pa. It should be emphasized that this positive value p_0 can be interpreted as "critical" pressure of the Universe, determined by means of dimensional analysis in terms of its critical density ρ_0. We specifically focus on this fact to point out a fundamental difference from the negative pressure of the Universe [8], which is associated with the phantom dark energy that may influence the effect of "accelerating" expansion of the Universe.

A natural explanation for the apparent accelerating expansion of the Universe and a few other important items follow from the consideration of the dimensionless parameter π_2 in the form

$$\pi_2 = \frac{kT}{mc^2} \sim 1$$

leading to the estimation of the particle mass m_0 for the media with the temperature T_0 and the disturbance propagation velocity c_0

$$m_0 \sim \frac{kT_0}{c_0^2} = 4.25 \cdot 10^{-40} \, \text{kg}.$$

An important conclusion of the analysis π_2 is directly proportional of the velocity c_0 to the square root of temperature $c_0 \sim \sqrt{T_0}$. In this connection we would like to get two remarks. The first is the possibility of a superluminal speed for disturbances in physical vacuum by $T>T_0$. In our case the limitation on week propagation perturbations velocity is withdraw and the superluminal motion (without violation of the causality principle) is allowed. Here relevantly one gives the analogy with the classic gas dynamics, which allows the motion with supercritical (supersonic) velocities. The second remark is concerned the apparent accelerating expansion of the Universe. The Universe in the early time periods was hotter, had a higher temperature T_0 and therefore higher value c_0 . This fact naturally explains the observed luminosity week effect of distant supernovae.

It should be noted that the combination of dimensionless parameters π_1 and π_2 leads to a state equation of an ideal gas. We write the dimensionless parameter

$$\pi_3 = \frac{\pi_1}{\pi_2} = \frac{p}{nkT} \sim 1,$$

where $n=\rho/m$ - the concentration of particles.

A more accurate assessment for m_0 [78-80] gives

$$m_0 = 5.6 \times 10^{-40} \, \text{kg}.$$

Further there is based this particle structure as a classic dipole with ultra-elementary charge near $q_0=10^{-28}$ C. Hence we can consider in detail a new complex elementary particle - Hidden Mass Boson (HMB). Dimensional analysis in the presence of a characteristic charge allows getting two character values: frequency (similar as Langmuir frequency in plasma) and Debye radius of screening. The last radius plays the key role in our study.

The dimensional analysis will be conducted, relying additionally on two parameters. Parameter π_4 written as

$$\pi_4 = \frac{G\rho}{\omega^2} \sim 1$$

allows us to introduce the characteristic gravitational frequency ω_0 , the characteristic time period $t_0 = 1/\omega_0$ and the characteristic linear dimension $L_0 = c_0 t_0$ of our Universe

$$\omega_0 \sim \sqrt{G\rho_0} = 0.82 \cdot 10^{-18} 1/s$$

The value of the gravitational frequency is an analog of the plasma frequency, which characterizes the electric displacement of the negative charge from a positively charged layer. The exact value ω_0 is written (by an analogy with the definition of the plasma frequency) with a coefficient of proportionality $\sqrt{4\pi}$, i.e. $\omega_0 = \sqrt{4\pi G\rho_0}$.

We also consider the dimensionless fifth parameter

$$\pi_5 = \frac{hc}{kTl} \sim 1$$

allowing an assessment of the characteristic length $l_0 \sim \frac{hc_0}{kT_0}$ = 5.3·10-3 m. This quantity can be interpreted as estimation of the mean free path of particles for the physical vacuum, which has temperature T_0.

4.2. Elements of thermodynamics and statistical mechanics

An interesting result of dimensional analysis is the presence of particles with mass m_0 in a free radiating space (the physical vacuum) with the equilibrium temperature T_0 and the disturbance velocity c_0. The zero law of thermodynamics allows introducing a temperature as the state parameter. The temperature T_0=2.725K characterizes of cosmic background microwave equilibrium radiation state. By using a simple gas kinetic approach we can find the refined value of the mass of these particles. The averaged kinetic energy of random motion of particles is

$$E = \frac{m_0 v_{av}^2}{2} = \frac{3}{2} kT_0 = m_0 \frac{3}{2} \frac{R_y}{m_0} \frac{T_0}{N_A} = \frac{9}{8} m_0 c_0^2,$$

where the $k = R_y / N_A$ Boltzmann constant, R_y - the universal gas constant, N_A - the Avogadro's number. From this relation we obtain for HMB

$$m_0 = \frac{4}{3} \frac{kT_0}{c_0^2} = 5.6 \cdot 10^{-40} \text{ kg} \cong 3 \cdot 10^{-4} \text{eV}.$$

We calculate the gas constant $R = R_y / m_0 N_A$ and specific heat capacity c_v and c_p in the assumption of ideal gas with adiabatic index κ=4/3:

$$R = \frac{k}{m_0} = 0.25 \cdot 10^{17} \text{J / kg K}, \ c_v = 0.75 \cdot 10^{17} \text{J / kg K}, \ c_p = c_v + R = 1 \cdot 10^{17} \text{J / kg K}.$$

Further, following the traditional thermodynamics of ideal gas for radiation medium, we can write the classical state equation

$$p = \rho RT$$

or

$$p = (\kappa - 1)\rho e,$$

where $e = c_v T$ - the specific internal energy.

An important next step is to postulate the structure of radiation particles, which allows explaining the large number of effects and phenomena observed in the physical vacuum. Following [79-80], we consider a whole electrically neutral particle in the form of a dipole consisting of two parts with positive and negative charge equaled to about $5 \cdot 10^{-29}$ Coulomb. Thus, we actually introduce the new mass medium of bosons - Hidden Mass Boson (HMB) and can apply to its analysis well developed methods of statistical mechanics and thermodynamics.

Let us explain in more detail, why the mass particle radiation of the physical vacuum is taken in our analysis in the form of the dipole, called hereafter by the hidden mass boson. First, in this case the issue of physical vacuum polarization is extremely clear. In an external electric field orientation of the HMBs takes on power lines of the electric field, partly compensating for the external field. Thus, we obtain a clear physical interpretation of the Maxwell's displacement current in free space. Further, the energy flux vector of the electromagnetic field - Umov – Pointing vector indicates the direction of the HMB polarization under the influence of an external electromagnetic field. In particular, when charging of the capacitor without insulator between the plates HMBs are moving from outer space in between the capacitor plate space, providing in this case, the displacement current. Another important process of electron-positron pair birth in the physical vacuum in the collision of two sufficiently intense electromagnetic pulse [81] should be interpreted as a break in a certain (sufficiently large) number of dipoles - HMB followed by concentration of their parts of the same sign of charge at the centers of the electron and positron under the influence of forces including non-electromagnetic nature (e.g., gas dynamics, gravity, etc.). When implementing this scenario, the HMBs will determine the mass of the birth of baryonic matter in the physical vacuum.

Substantiated to some extent chosen postulated HMB structure, we proceed to the methods of statistical mechanics. We consider the important question of the consistency of our work findings to well-known basics and conclusions of statistical physics and thermodynamics of a boson gas and its special case - a photon gas with zero rest mass of photons. Below we demonstrate the absence of such contradictions.

First of all, we write the state equation for gases obeying Bose statistics and Fermi statistics. Here, the dimensionless parameter can be written [82, 83]

$$\pi_3 = \frac{p}{nkT} = 1 \mp \frac{1}{2^{5/2}} \lambda_0 + \dots,$$

where the signs $\mp$, respectively, for the Bose and Fermi gases, as well $\lambda_0^{-1} = (2\pi mkT)^{3/2} / h^3 n$ - the statistical sum per particle. The second and higher terms in the right side of the above formula are quantum statistical origin. This series expansion is valid for values $\lambda_0<1$ or to values $n<(2\pi mkT)^{3/2} / h^3$. The magnitude $h / (2\pi mkT)^{1/2}$ is the length of the thermal de Broglie wavelength, and the value $h / (3mkT)^{1/2}$ corresponds to the thermal energy of the particle *3/2kT*. Thus the Bose statistics in the usual manner is applied to the entered us with dimensional analysis and thermodynamics of the HMB gas. For future study the condensation state of the HMB gas on the lowest temperature and high pressure is also very important item.

As for the well-known theory of blackbody radiation, in our approximation it requires some revision, which consists in passing from the linearized formulation of the problem to a fully nonlinear formulation. The essence of this transition can be clearly illustrated by the linearized theory of acoustics and the original theory of the nonlinear gas dynamics. It seems very clear that in a limited gas volume it is possible to describe the it's state in the acoustic approximation using the equation for the acoustic pressure perturbation

$$\nabla^2 p - \frac{1}{c^2}\frac{\partial^2 p}{\partial t^2} = 0$$

Based on this description, we can then determine an acoustic sound radiation from a closed cavity and the radiation field in the outer vicinity of the limit volume. In this description one can calculate the acoustic field in a gas without going into details of what the gas medium consists of a mass of individual particles. However, to calculate the real total pressure in the limited volume the initial nonlinear laws of gas dynamics are required (of them will be discussed in the section 4.4).

A similar situation occurs with the thermal radiation of a blackbody. Currently, in the approximation of massless photon equilibrium radiation field is described by the linear wave equation [82]

$$\nabla^2 E - \frac{1}{c^2}\frac{\partial^2 E}{\partial t^2} = 0.$$

The radiation field is represented by a set of simple harmonic oscillators with discrete spectrum of energy. As a result of the energy spectrum of equilibrium radiation gives the Planck formula

$$U_\nu = \frac{8\pi h\nu^3}{c^3}\frac{1}{e^{\frac{h\nu}{kT}} - 1}$$

Calculations using this formula are in full agreement with the experimental far field radiation, where the linearized theory actually works (as in acoustics case). However, in the case of massive bosons radiation the study of the thermal process in limited volume should be with help of nonlinear formulation, which we detail below.

To conclude this section we emphasize that known values p_0 and ρ_0 characterize the state of the photon gas in the physical vacuum of space (in the vicinity of the Earth and solar system) with the measured temperature CMB $T_0 = 2{,}725\ K$. The pressure p_0 is the direct pressure of the HMBs medium in the region (source) at T_0. This value differs on a factor $c_0=3\cdot10^8$ m / s from the linearized values in a far field, which is calculated by methods of classical electrodynamics. It should specify the complete analogy with the usual gaseous medium in which acoustic pressure away from the noise source is different from the pressure at the source (this difference is also characterized by a factor containing the magnitude of the velocity of disturbance propagation, in this case - the sound speed c).

Further consideration of the influence of thermal radiation effects environment will be carried out within the framework of two-component model of the emitting gas with the mass of the photon component [84].

4.3. Modified Maxwell equations in free space

In the medium of physical vacuum being examined the longitudinal and transverse waves will propagate in the electromagnetic field. The system of linear equations, describing the propagation of electromagnetic disturbances can be written in the form [40]

$$\begin{aligned}
&\frac{\partial \boldsymbol{E}}{\partial t} - c\ rot\boldsymbol{H} + c\ gradq_e = 0,\\
&\frac{\partial \boldsymbol{H}}{\partial t} + c\ rot\boldsymbol{E} + c\ gradq_m = 0,\\
&\frac{\partial q_e}{\partial t} + c\ div\boldsymbol{E} = 0,\\
&\frac{\partial q_m}{\partial t} + c\ div\boldsymbol{H} = 0.
\end{aligned}$$

With the purpose of longitudinal waves modeling into tradition system of electrodynamics equations of a free space, the scalar field q_e and q_m were introduced and that represents the density of force lines of electric and magnetic field. The first term in the first equation of the system is a time change of E and the displacement current in vacuum.

The system describes the propagation of all parameters of disturbances with the same speed c. It is easy to prove, differentiating the first two equations with respect to t:

$$\begin{aligned}
&\frac{\partial^2 E}{\partial t^2} + c^2 rot\ rotE - c^2 grad\ divE = 0,\\
&\frac{\partial^2 H}{\partial t^2} + c^2 rot\ rotH - c^2 grad\ divH = 0,
\end{aligned}$$

that, by using the formula of vector analysis

$$\Delta A = grad\ divA - rot\ rotA$$

leads to the waves equations for the changes E and H

$$\frac{\partial^2 E}{\partial t^2} - c^2 \Delta E = 0, \quad \frac{\partial^2 H}{\partial t^2} - c^2 \Delta H = 0.$$

The changes of scalar field q_e and q_m , as well as scalar and vector potentials of E and H can be also lead to the wave equations.

The system of linear equations describes the propagation with equal velocities both the longitudinal waves (the wave of compression and rarefaction) and the transverse wave (share wave) in a homogeneous isotropic space. At the same time, the modeling of longitudinal wave is provided by the introduction additionally of the scalar fields q_e and q_m to the system Maxwell equations. In this connection, we can call the scalar fields q_e and q_m as fields of density (force lines), connecting with fields E and H. In principle, it is not difficult to examine the case when propagation velocities of longitudinal and transverse waves in the electromagnetic field are different.

4.4. Conservation laws for two component gaseous medium

Here is a complete system of conservation laws for two component model of gas-like environment, taking into account the HMB medium (radiation) component. All used parameters will be denoted in the traditional way. Attributing them to the corresponding indices: g - for the gas component, f - for the radiation component (e.g. density ρ_g and ρ_f).

The total value of the density, pressure and internal energy will be denoted without an index.

The laws of conservation of mass, momentum and energy in the divergence form for each component have the form [84]

$$\begin{aligned}
&\frac{\partial \rho_g}{\partial t} + div(\rho_g \bar{V}_g) = q_g, \\
&\frac{\partial \rho_f}{\partial t} + div(\rho_f \bar{V}_f) = q_f, \\
&\frac{\partial \rho_g \bar{V}_g}{\partial t} + div(\rho_g \bar{V}_g (\bar{V}_g \cdot \bar{n})) + grad\ p_g = r_g, \\
&\frac{\partial \rho_f \bar{V}_f}{\partial t} + div(\rho_f \bar{V}_f (\bar{V}_f \cdot \bar{n})) + grad\ p_f = r_f, \\
&\frac{\partial \rho_g e_g}{\partial t} + div(\rho_g e_g \bar{V}_g) + p_g div \bar{V}_g = div(K_g grad T_g) + c_{fg}(T_f - T_g) + Q_g\ , \\
&\frac{\partial \rho_f e_f}{\partial t} + div(\rho_f e_f \bar{V}_f) + p_f div \bar{V}_f = div(K_f grad T_f) + c_{fg}(T_g - T_f) + Q_f
\end{aligned} \tag{1}$$

This system of equations is written for the thermal conductivity of the gas and radiation components (the first terms on the right hand side, K_g and K_f - the thermal diffusivity of the gas and radiation components, respectively). The second terms on the right sides of two last equations characterize the energy exchange between the gas and radiation components. The last terms, Q_g and Q_f are supplementary sources of energy, taking into account the availability of additional channels of energy exchange (e.g., in the case of registration of chemical reactions, etc.). The system (1) is closed by equations of state for the gas and radiation components. Using this set, one can give a natural interpretation of the growth of entropy in compression shocks, where a part of the kinetic energy of the gaseous component in the shock turns into heat. When the velocity V_g drastically decreases, a part of the energy from polarized molecule space goes into the radiation component and this part of energy is dissipated due to heat conduction.

Of course, a solution of this system in general form involves considerable difficulties, because we need to specify the value of the exchange heat coefficient c_{fg} between the phases. Substantial simplification can be achieved by considering the approximation of one velocity and one temperature movement phase in the presence of a thermodynamically equilibrium

$$V_g=V_f=V; \quad T_g=T_f=T.$$

We also assume that there are no external sources of mass and momentum in this region of flow and mass transfer between phases:

$$q_g=q_f=0; \quad r_g=r_f=0.$$

Then, following [85, 86], we represent the continuity equation of each phase in the form

$$\frac{1}{\rho_g}\frac{d\rho_g}{dt}+div\bar{V}=0,$$

$$\frac{1}{\rho_f}\frac{d\rho_f}{dt}+div\bar{V}=0$$

or

$$\frac{d}{dt}\left(\ln\frac{\rho_f}{\rho_g}\right)=0.$$

The last equality says the preservation of values

$$\alpha=\rho_f/\rho_g \tag{2}$$

along the stream lines, and if we assume that the initial time the density ratio is constant and independent of the coordinates, the equation (2) is valid at any point of considered medium.

We write total conservation laws for both components of the medium. Adding first two and second two equations in the system (1) we obtain, taking into account our assumptions, usual equations of a continuity and a motion for one-component medium

$$\frac{\partial \rho}{\partial t} + div(\rho\, \bar{V}) = 0,$$
$$\frac{\partial \rho\, \bar{V}}{\partial t} + div(\rho \bar{V}(\bar{V} \cdot \bar{n})) + grad\ p = 0\ .$$

Slightly change the total energy equation. Adding the last two equations in (1) we obtain

$$\frac{\partial}{\partial t}\left[\left(c_{vg}\rho_g + c_{vf}\rho_f\right)T\right] + div\left[\left(c_{vg}\rho_g + c_{vf}\rho_f\right)T\right] + pdiv\bar{V} = -divW + Q,$$
$$-W = K_g gradT_g + K_f gradT_f,\ Q = Q_g + Q_f.$$

In order to give the energy equation the usual form, we transform the expression

$$N = c_{vg}\rho_g + c_{vf}\rho_f,$$

as

$$N = c_{vg}\rho_g(1 + (c_{vf}\rho_f)/(c_{vg}\rho_g)) = c_{vg}\rho/(1+\alpha)\left(1 + \left(k_g - 1\right)/\left(k_f - 1\right)p_f/p_g\right) = \bar{c}_v\rho,$$
$$\bar{c}_v = c_{vg}\left(1 + \left(k_g - 1\right)/\left(k_f - 1\right)p_f/p_g\right)/(1+\alpha),\ k = c_p/c_v.$$

Recall that, according to (2) $\alpha=\rho_f/\rho_g$, constant along streamlines. Consequently, when $T_f=T_g$, the ratio p_f/p_g is constant along streamlines.

Similarly, we transform the equation of state for the total system

$$p = \bar{R}rT, \quad \bar{R} = R_g\left(1 + p_f/p_g\right)/(1+a).$$

The energy equation with right term *-divS*, where the radiation flux $S=\sigma T^4$, is often used in the simulation of the radiate flows. For further analysis with using of this approximation we obtain the following system, describing the equilibrium one velocity flow in the presence of radiation effects

$$\frac{\partial \rho}{\partial t} + div(\rho\, \bar{V}) = 0,$$
$$\frac{\partial \rho\, \bar{V}}{\partial t} + div(\rho \bar{V}(\bar{V} \cdot \bar{n})) + grad\ p = 0\ ,$$
$$\frac{\partial}{\partial t}\bar{c}_v T + div\bar{c}_v T + pdiv\bar{V} = -div\sigma T^4,$$
$$p = \bar{R}\rho \mathrm{T}.$$

Obviously, when $\rho_f=0$ we obtain the ordinary system of gas dynamics equations for one component radiation medium. Further we present some examples of our model applications.

5. Space energy & grand unified theory

"Space Energy is conserving,
Space Entropy is growing".
E. Fermi

A classical unified theory of electromagnetic, weak and strong interaction, electrovalence linkages and antimatter is built coming from our Space Energy simulation. The theory bases on Hidden Mass Boson in form of a classic dipole and the Debye length of polarized spaces. These items are the background of our Grand Unified Theory (GUT).

5.1. Debye spheres of electron and proton – the background of GUT

"The electron polarizes vacuum, allegedly attracts to itself virtual positrons, and repulses virtual electrons". So, in particular, the vacuum polarization phenomena are described in monograph [89]. In full correspondence with similar experimentally confirmed polarization of near electron space is found the stated below mathematical model of electron and proton structure. Here we shall imply only under "virtual" electron and positron a real polarized gaseous structure of DM dipole particles (HMBs), from which, as already noted, electron-positron pairs might be generated.

Within the framework of considered approximation one can formulate specific equation system similar the gas dynamics model with electric charge presence, as the two fluid hydrodynamic plasma model. For steady state from mentioned models we can lightly derive equations, describing particle electricity potential and concentration distribution in polarized Debye spaces of electron, positron, proton and anti-proton.

First of all we would like to obtain the estimation of Debye spheres for electron. A Debye sphere is a volume of influence, outside of which charges are sufficiently screened. In out simulation a Debye sphere is formed with a pressure field presence of HMB gaseous medium. For steady state a two fluid hydrodynamic model [90, 91] may be expressed by equation system (in the SI units)

$$\begin{aligned} &\frac{\nabla p_{\pm}}{n_{\pm}m} = \mp\frac{1}{4\pi\varepsilon_0}\frac{e}{m}\nabla\varphi, \\ &\Delta\varphi = -4\pi e(n_{+} - n_{-}), \\ &p_{\pm} = n_{\pm}kT, \end{aligned} \tag{3}$$

where $p_{\pm}, n_{\pm}$ - pressure and concentration of positive and negative of HMB components, which have the same temperature T, e – elementary charge of HMB dipoles. From (3) the equation of electric potential φ for polarized space in non-dimensional form is written in rather simple form

$$D^2\Delta\varphi = 2sh\varphi. \tag{4}$$

Here potential φ is referred to its value $\varphi_0 = 4\pi\varepsilon_0 T$, $D = sqrt(\varepsilon_0 kT / q^2 n_0)$ - the Debye radius (length), n_0 - typical HMB boundary concentration. For concentration of HMB particle charge component in polarized space we have from (3) the Boltzmann distribution (in non-dimensional form)

$$n_{+,-} = n_0 \exp(\mp\varphi)$$

The Debye length for electron in HMB medium can be calculated using its mass m_e=9.1·10^{-31}kg, charge q_e=1.6·10^{-19}C and Compton's length λ_e=2.43·10^{-12}m. The effective electron volume, which contains the most part of its mass, may be estimated as

$$V_e = 4/3 \cdot \pi \cdot \lambda_e^3 = 0.601 \cdot 10^{-34}\ \text{m}^3.$$

Our great assumption is the distribution of whole electron mass between the point center (with the smallest point radius near 10^{-17} -10^{-18} m, where the whole electron charge is concentrated) and other volume V_e. Here we assume that the electron mass is dividing near half. When HMB averaged concentration n_e in this volume is

$$n_e = (m_e/2)/V_e/m = 1.35 \cdot 10^{43}\ 1/\text{m}^3.$$

Further we provide the estimation of HMB dipole charge q. The number of charge particles in the electron center N_e=(m_e/2)/(m/2)=1.62·10^9 and q=q_e/N_e=0.99·10^{-28}C. Now we can calculate the Debye electron length of charge screening (by T=2.725K)

$$D_e = sqrt(\varepsilon_0 kT/q^2/n_e) = 0.5 \cdot 10^{-10}\text{m}.$$

In the spherical symmetric case with single spatial coordinate-radius r we come from (4) to relationship

$$\frac{D_e^2}{r^2}\frac{d}{dr}\left(r^2\frac{d\varphi}{dr}\right) = 2sh\ \varphi$$

We shall bring typical solutions of the last equation for polarized electron space (Figure 4). A principal important particularity of distribution presented is potential pit and barrier on external border of polarized space with distribution $\varphi(r)$ break. Herewith an induced negative electric charge is concentrated on external border, induced by charged electron core during polarizations of its "fur coat". The results presented are valid due to charge symmetry also for description of polarized positron space structure (when changing potential sign on opposite). The scheme of polarized electron space is shown on Figure 5.

Now consider possible building on the same principles of proton and antiproton model. In analogy with electron we suppose that the entire positive proton charge is concentrated in its center, having an estimated size, like electron core size of order 10^{-17} m. This central nucleus is spherical "droplet" surrounded polarized by HMB in "fluid" aggregate condition. The droplet size defines a typical known proton nuclear size (about $0.8 \cdot 10^{-15}$ m), where practically its entire mass concentrates. The concentration of HMB in the fluid droplet is near 10^{57} and the Debye screening radius near $0.8{\cdot}10^{-15}$m. The HMB fluid droplet provides the effective Debye screening and plays the key role in week and strong interactions. Around proton nuclear droplet there is polarized spherical HMB space in gaseous aggregate condition, similar to polarized positron space. Modeling structure of two-layer polarized proton space is executed by integrating equation (4) for electric potential with the use of different state equations (for liquid and gaseous HMB phases). Examples of similar solutions by nature repeat solutions shown on Figure 4. And again basically important here is the presence of two potential pits and two barriers at $r \approx 0.8 \cdot 10^{-15}$ m and $r_0 = D_e \approx 5 \cdot 10^{-11}$ m. The structure built of proton with partial screening of its charge by two-layer polarized space is steady. Due to charge symmetry the antiproton structure repeats proton structure (with corresponding change of potential sign and others). The scheme of polarized proton space is shown on Figure 6.

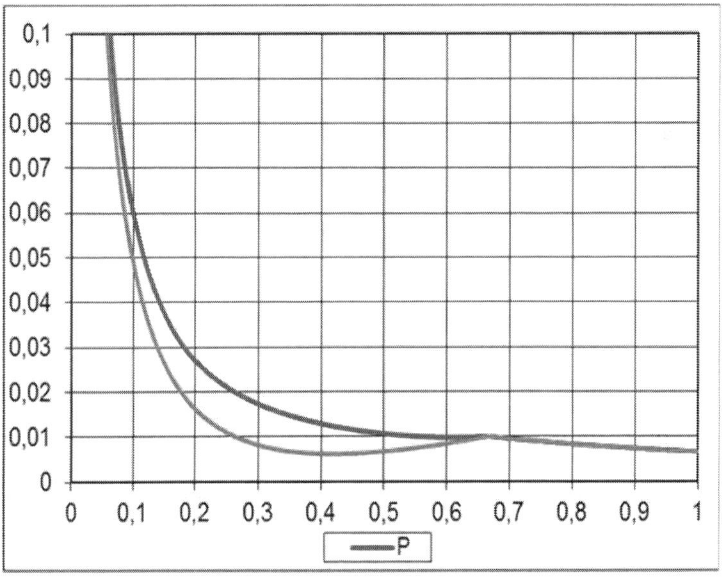

Figure 4. Potential distribution in electron screening space for two D

The model of hydrogen atom consists of proton with concentrated in its center positive charge and two-layer HMB polarized space in fluid (up to $0.8{\cdot}10^{-15}$m) and gaseous aggregate conditions and having an electron situated in steady stationary state on polarized space external surface ($r_0{=}D_e{=}0.5{\cdot}10^{-10}$). Herewith for ensuring electron stationary condition in atom there is no need to introduce additional quantum postulates. This condition is provided by potential distribution in the electron and proton models built. The scheme of polarized hydrogen atom space is shown on Figure 7. The potential distribution in the hydrogen atom is present on Figure 8.

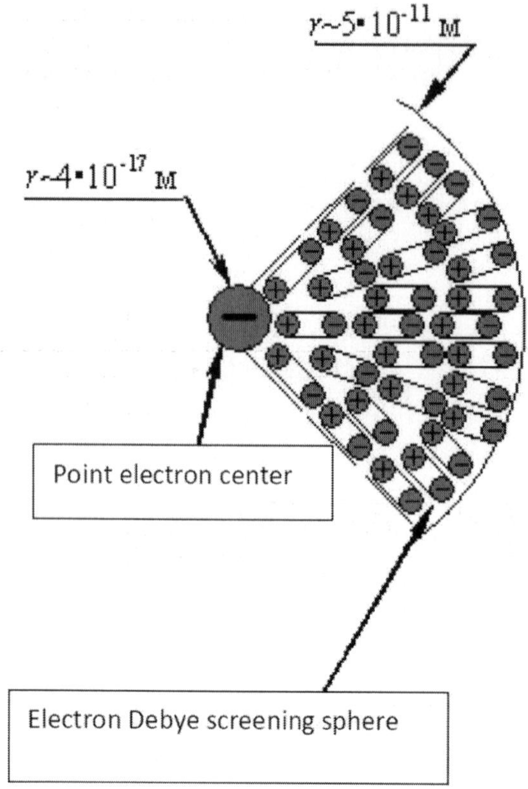

Figure 5. Scheme of electron screening space

5.2. Physical model of van der Waals polarized atom spheres

The presence of real existing polarized atom space and their forms close to spherical (van der Waals spheres as Debye screened spheres) is validated by methods of crystal-chemistry and scanning probe microscopy. By means of tunnel microscope needle one can manipulate separate spherical atoms, move them on surfaces, build 3D figures, etc. Atom shape presentation as van der Waals spheres is now broadly used method when studying substance physical-chemical properties.

The previous section offers models of polarized screened spaces partly shielding central concentrated electrical charges that enable demonstrative physical interpretation of van der Waals atom spheres. These spheres are in essence external borders of polarized spaces filled with HMB dipoles Coulomb interaction bound. On such spherical borders with local concentrated induced charges might be stationary placed concentrated charges of another sign (example: hydrogen atom from article previous paragraph).

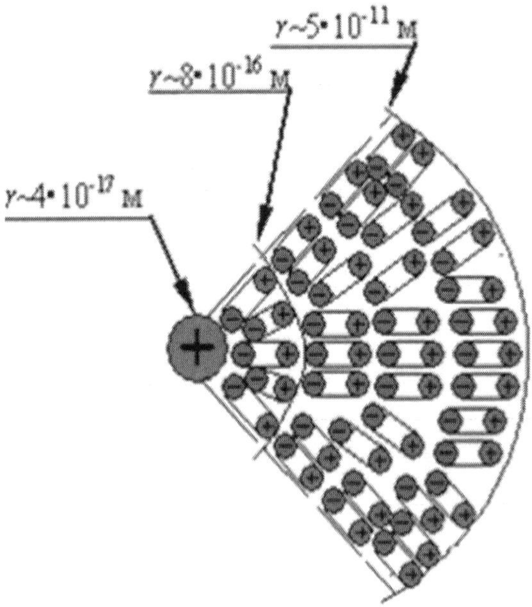

Figure 6. Scheme of proton screening two layer space

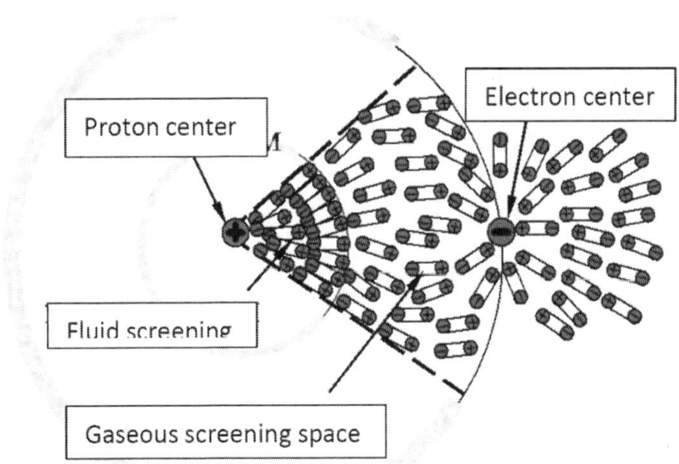

Figure 7. Scheme of hydrogen atom

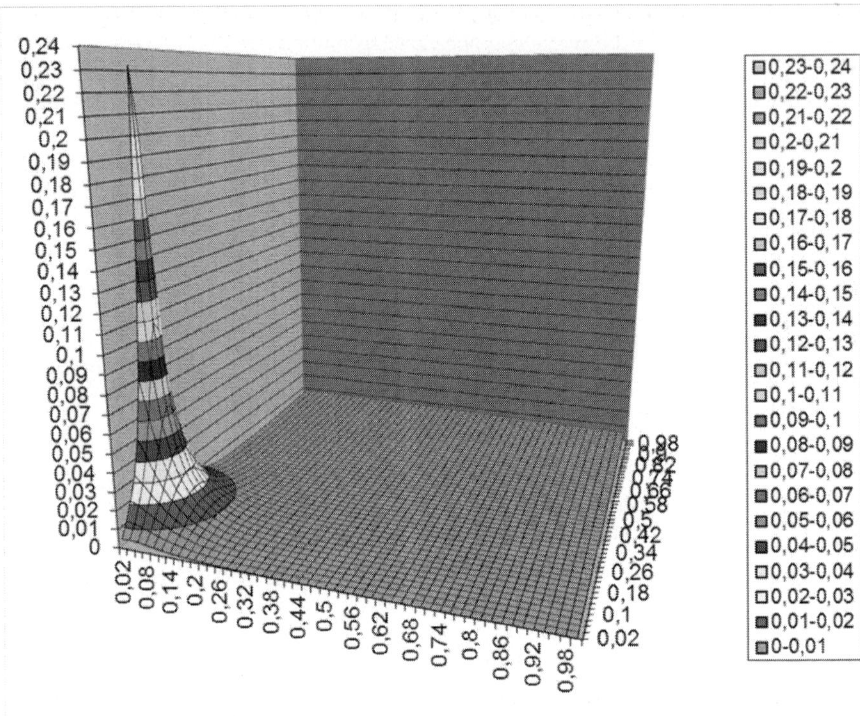

Figure 8. Numerical solution for potential distribution inside hydrogen atom

Possible stationary steady electron position on van der Waals spheres brings about a model of atom electronic shell with STationary ELectron (STEL models). Electronic shells of spherical shape are equidistant to nucleus centre at distances accordingly $1r_0, 2r_0, 3r_0, 4r_0$, etc. (natural row coefficients 1, 2, 3, 4... show, as accepted, main quantum number). The first shell area $4\pi r_0^2$ is fully occupied by 2 stationary electrons; they rather tightly cover nearly whole area by their own polarized spaces. The following three shells area $4\left(4\pi r_0^2\right)$, $9\left(4\pi r_0^2\right)$ and $16\left(4\pi r_0^2\right)$ can be occupied by less 8, 18, and 32 stationary electrons (double square of main quantum number); also tightly covering by their own polarized spaces nearly whole disposable area of corresponding shell. The rule stated defines maximum possible electrons number on any electronic shell (its "thick packing"), however electrons number on external shells at once may not reach its maximum value.

Number of ordinal atom element increasing and its nucleus positive charge growing under action of increasing Coulomb forces, the electron shells compress (r_0 decreases) that fully complies with experimental data on atom sizes of different chemical elements. Heat expansion of electronic shells (including external van der Waals spheres) on linear law is described by corresponding solutions change of equation (4) for polarized atom space when changing Debye radius $D = (kT\varepsilon_0 / n_0 q^2)^{1/2} \sim T$ (since typical gaseous HMB concentration n_0 at similar pressure p_0 is inversely proportional to temperature T).

5.3. STEL model for atoms and molecules

Let us give some examples of typical atomic structures within the framework of STationary ELectron (STEL) model. Begin with helium atom and its first electronic shell filled by two electrons. There are two protons and two neutrons in the center of helium atom. On surfaces of first polarized space – helium nucleus border ($r \sim 10^{-15}$ m) there are two electrons in steady state due to total potential distribution. The DM polarized space in gaseous condition is situated around nucleus on the space external shell ($r \sim 0.5 \cdot 10^{-10}$ m) with two electrons in steady state. This first electronic shell is tightly "filled" due to the described structures of electron polarized space (also $r \sim 0.5 \cdot 10^{-10}$ m). Each electron occupies about half spherical shell around helium atom.

The next is a carbon atom. Its external shell has 4 electrons located in tetrahedral tops (traditional direction of valence links, Figure 9). Oxygen atom on external shell contains 6 electrons. The least total potential energy of these electrons is provided by their location in two triangles tops in parallel planes and turned to each other by angle 60° (so called triangle anti-prism, Figure 9).

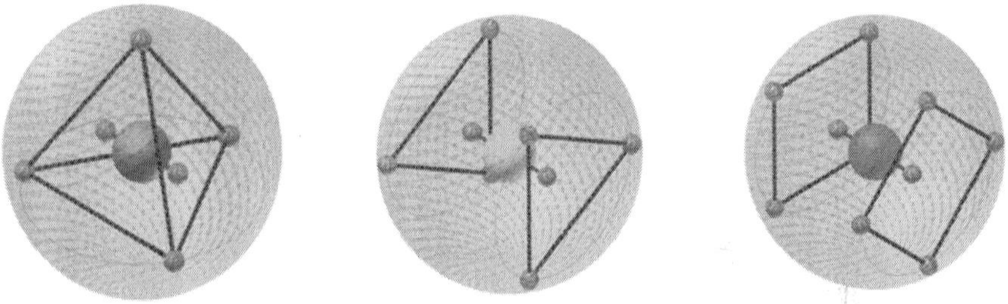

Figure 9. Structures for carbon, oxygen, and neon atoms

An inert gas neon atom has on external shell 8 electrons. The least total potential energy of these electrons forms at their location in square tops turned by angle 45° (square anti-prism, Figure 9). Similar structure to carbon, oxygen, and neon atoms is for atoms of silicon, sulphur, and argon accordingly with the same location of 4, 6, and 8 electrons on the third external shell. One can uncomplicated similar way reproduce stationary electron location on completely filled third shell with 18 electrons of 4 period (5 rows) elements and fourth shell with 32 electrons of 6 period elements of the Mendeleyev table.

The illustrated application of STEL model for electron shells description in some features repeats the known Lewis octets. An important novelty to model STEL is electrons location in anti-prism tops (for oxygen, neon, sulphur, and argon) providing minimum value of total potential energy of external electrons.

Existence of DM polarized spaces of electron, proton, and atomic nucleus allows description from general positions through Coulomb interaction of any available molecular linkages as stationary deterministic structures of shell electrons and nucleus. Here we also use the model STEL. The simplest hydrogen H_2 molecule presents a 2D quadrangle, in opposite tops of which there are two protons and two electrons (Figure 10). These perimeter concentrated charges are located at distances about Debye radius and partly shielded by HMB polarized space. Potential distribution in polarized hydrogen molecule space might also be found by integrating equation (4). Here the covalence link presents a pair of stationary electrons (not moving on orbits).

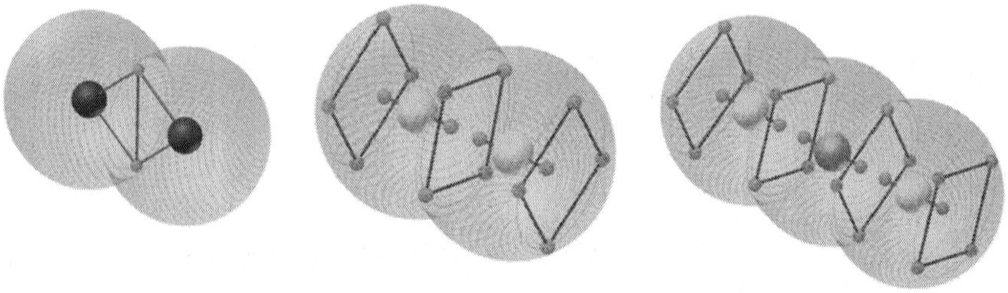

Figure 10. Structures for H_2, O_2 and CO_2 molecules

For geometric presentation and physical description of molecular links with participation of elements of second and third periods of Mendeleyev table we return to modified Lewis cubic structures and his electron octets. Instead of perfect cubes with electrons in their tops we have square anti-prisms. Bring some typical examples. In asymmetrical water molecule each hydrogen atom is connected by covalence link (pair of socialized electrons) with oxygen atom in neighboring tops of square anti-prism.

Oxygen molecule is united by double covalence link. We have here on the second orbital shells of each atom electron octets in anti-prism tops (Figure 10). An example of "twisted" molecule with double covalence link serves the molecule of carbon dioxide (Figure 10).

Similar geometric models with square anti-prism structure (electron octets) and covalence links are typical for carbon compounds. Show, for example, saturated hydrocarbon with filled octet of the electronic shell - methane (Figure 11). Compounds, similar to carbon, of third period silicon element with four valency electrons on the following third electronic shell are also comfortable to present in the form of geometric models with square anti-prisms; with the only difference - in covalence links participate electrons of silicon third external electronic shell.

Within the framework of STEL model it is natural ("classical") way to represent models of hydrogen links, when on the first, nearest to proton (hydrogen ion) shell are stationary localized two electrons simultaneously belonging to two negative ions. For instance, hydrogen difluoride HF_2^- possesses linear structure $F^-H^-F^-$; its steadiness is obliged to the proton linking two negative fluorine ions by Coulomb interaction (Figure11).

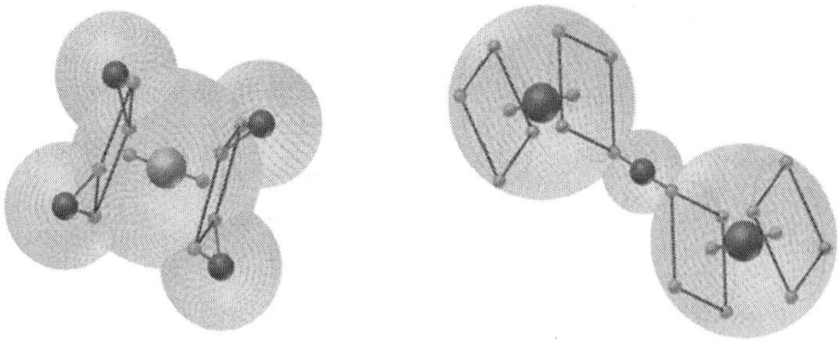

Figure 11. Structures for CH_4 and HF_2 molecules

The ionic link presents Coulomb ion interaction when external electrons of one atom complement more filled external shell of another atom. Here it is also comfortable for elements of second and third periods of the Table to consider electron octets of external shells with geometric approach in the form of anti-prisms. Typical examples serve sodium chloride and sulphureous magnesium.

Now describe from used deterministic positions and STEL model metallic links when atom external shells are partly filled by electrons. Metallic links are realized through links of electronic shells nearest to the external shell, through one or several electrons, which begin to belong simultaneously to two nearby atoms. Then electrons of external blank shells and "spare" electrons of binding shell (if and when before links this shell was filled) become not connected with concrete atom and easy moving. Examples are schematic traditional 2D presentations of metallic electronic links for lithium and natrium (Figure 12). For nickel subsequent to the second filled electronic shell with 8 electrons octet follows the third shell with 16 electrons that is complemented up to 18 electrons by attaching nearby atom two electrons realizing metallic link.

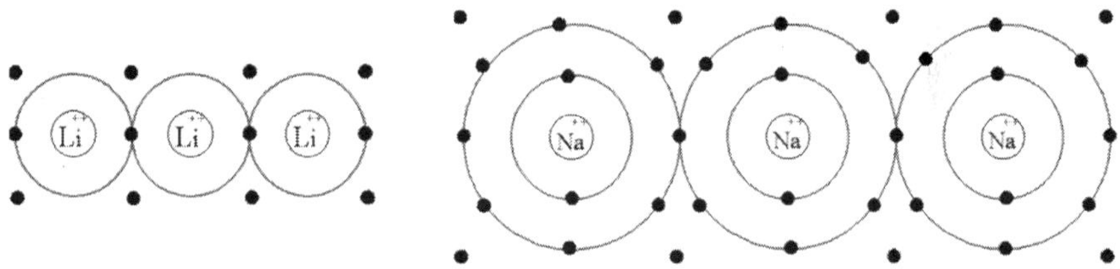

Figure 12. Schemes of metal linkages for Li and Na

Special consideration has fulfilled for hydrocarbon valence linkages in benzoyl C_6H_6 and naphthalene $C_{10}H_8$ molecules. We give the molecule images, their detail structures (Figure 13) and also electronic and structural formulas (Figure 14). It should be emphasized that the benzoyl ring has internal linkages between carbon atoms through three stationary electrons.

The next examples are related to borohydrogen molecules B_2H_2, B_6H_6, B_4H_{10} (Figure 15). The complex B_2H_2 has special type of internal four electrons linkage and outer four valence linkages, similar the carbon and silicon atoms.

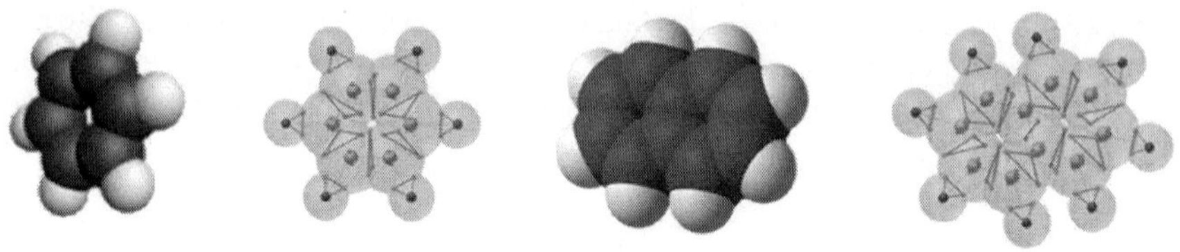

Figure 13. Benzoyl C_6H_6 and naphthalene $C_{10}H_8$ molecules and their form

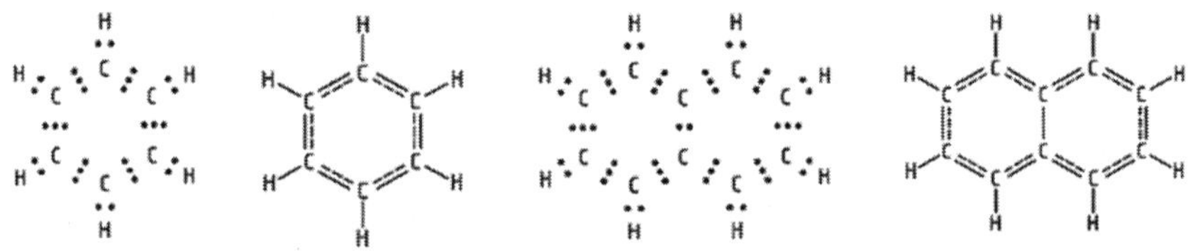

Figure 14. Electronic and structural formulas of benzoyl C_6H_6 and naphthalene $C_{10}H_8$ molecules

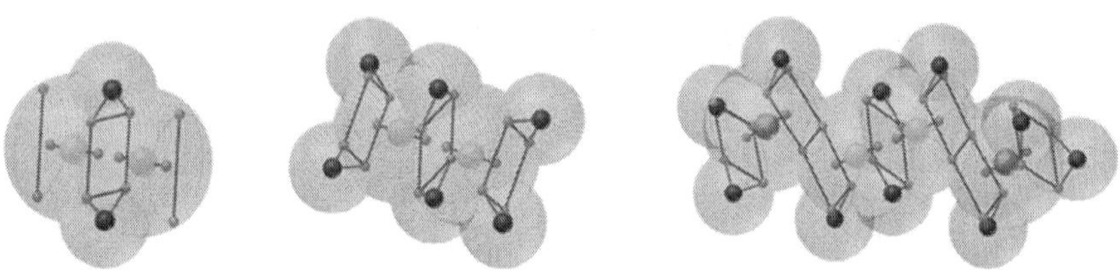

Figure 15. Complex $B_2 H_2$ and molecules $B_2 H_6$ and $B_4 H_{10}$

In the approach presented any molecular links are realized by stationary electrons of external orbits by means of electric (Coulomb) interaction with partial screening electron and nucleus "point" charges by DM polarized particles. Electric potential distribution is described in this approach by the same equation (4) with stationary electrons distribution providing minimum of total potential energy. The principle of minimum total potential energy brings about electrons stationary localization in geometric anti-prism tops that provides maximum repulsion of electrons on electronic shells. Anti-prism foundations turning are a base for shaping heliciform circuits of complex molecules (in particular, RNA and DNA molecules).

It follows to emphasize in the section conclusion that stated physical models allow looking from new positions at known ways of modeling chemical links. These methods may include the Lewis model and diagrams and Van der Waals sphere model for revealing intermolecular contacts.

5.4. Examples of weak and strong interactions

Following our methodology we show now shortly the scheme of neutron as a positive nuclear of proton with Debye screening (up to $r_0=0.8\cdot10^{-15}m$) by liquid layer of HMB dipoles and stationary electron, presenting on boundary of Debye region (Figure 16). Decay of neutron gives naturally proton, electron and antineutrino, which presents itself isolated soliton in DM gaseous medium. Exact soliton solutions for neutrino and antineutrino were obtained in papers [92-95].

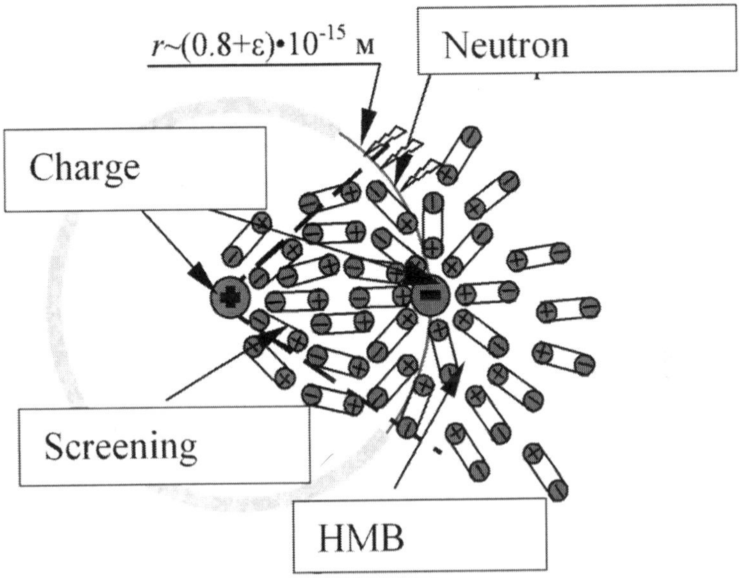

Figure 16. Scheme of neutron

The STEL model with solution (4) allows us to simulate also an internal nuclear structure. Figure 17 shows the typical nuclear structures for deuterium, tritium, helium ^{3}He and ^{4}He. Potential distribution for deuterium was calculated (Figure 18). Figure 19 present the nuclear structure of ^{4}Li, ^{5}Li and ^{6}Li.

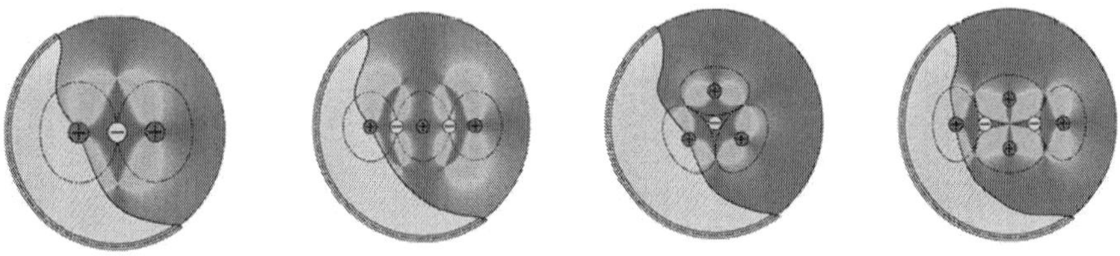

Figure 17. Internal structures of deuterium, tritium and helium ^{3}He and ^{4}He

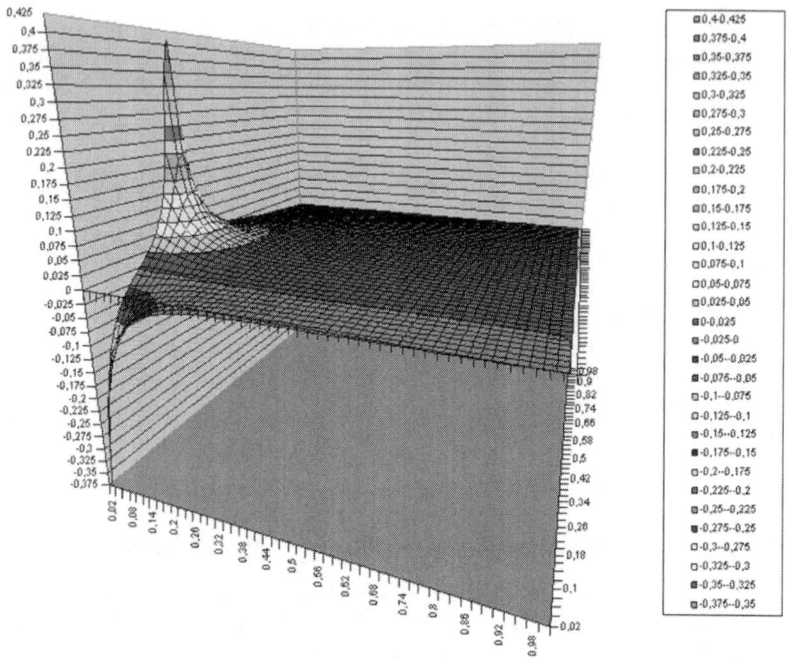

Figure 18. Numerical solution for potential distribution inside deuterium

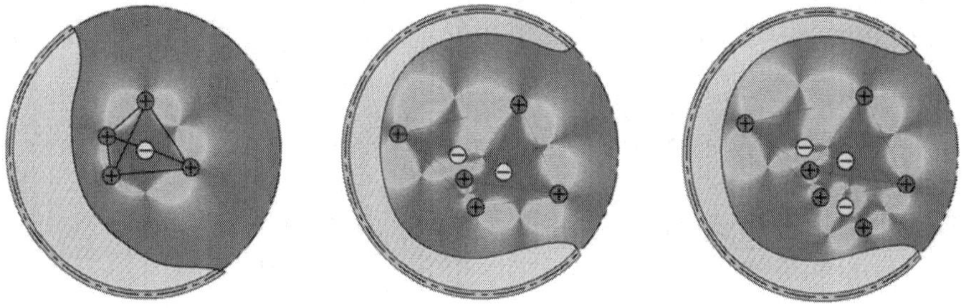

Figure 19. Internal structures of Lithium ^4Li, ^5Li and ^6Li

6. Additional experimental confirmations

"Experience is and a source of knowledge and a criterion of truth.".
Francis Bacon (1620)

In the last section of the chapter we would like to demonstrate additional experimental confirmations of our simulation and first of all the HMB (DM) medium presence in Cosmic space and on Earth. Here, as good examples, one can select wide experimental data on solid bodies, deoxyribonucleic acid (DNA) molecules, some aeronautics and astronautics.

6.1. Hooke's, Dulong–Petit's and thermal expansion laws

The solid body is characterized by structural rigidity and resistance to changes of shape or volume. The atoms in a solid are tightly bound to each other, either in a regular geometric lattice (crystalline solids) or irregularly (an amorphous solid). The forces between the atoms in a solid can take a variety of forms but today we take into account mainly electrostatic bonds. For example, a crystal of sodium chloride (common salt) is made up of ionic sodium and chlorine, which are held together by ionic bonds. In diamond or silicon, the atoms share electrons and form covalent bonds. In metals, electrons are shared in metallic bonding. Some solids, particularly most organic compounds, are held together with van der Waals forces resulting from the polarization of the electronic charge cloud on each molecule. The dissimilarities between the types of solid result from the differences between their bonding. New our item is the presence inside bodies (similar by containers) HMB (DM) medium with own pressure, which play the same important role also as electrostatic bonds and steady solid forms result by equilibrium between HMB medium pressure and electrostatic atomic forces.

Hooke's law. Hooke's law was found in 1610. The most commonly encountered form of this law is probably the spring equation, which relates the force exerted by a spring to the distance it is stretched by a spring constant k, measured in force per length

$F = -k\Delta x$. The negative sign indicates that the force exerted by the spring is in direct opposition to the direction of displacement. It is called a "restoring force", as it tends to restore the system to equilibrium. In equilibrium state electrostatic forces and internal HMB medium pressure forces equal (the first equation in the system (3)). HMB pressure is increasing (or decreasing) proportional of displacement x (as for ideal gas case).

The potential energy stored in a spring is given by $P_e = k\Delta x^2/2$, which comes from adding up the energy it takes to incrementally compress the spring. That is, the integral of force over displacement. (Note that potential energy of a spring is always non-negative.) This energy is internal energy of HMB gaseous medium and also can be calculated using our gas dynamics modeling.

For HMB gaseous medium we have the state equation $pV=NkT$. When $T=const$ and $N=const$ the value $pV=const$. For the pressure increasing Δp with the change volume $\Delta V=V(1-\Delta x/l)$ one can be written

$$(p+\Delta p)V(1-\Delta x/l)=NkT$$

and

$$\Delta p \approx \Delta x/l\ nkT.$$

Hence Young's modulus

$$E=\Delta p/\Delta x/l \approx nkT.$$

As an example we determine the concentration n on boundary polarized space for aluminum with $E=70GPa$ and the Debye temperature $T=394\ K$

$$n \approx E/kT = 1.3 \oplus 10^{31}\ 1/m^3.$$

Dulong–Petit's law. The Dulong–Petit law, a chemical law proposed in 1819 by French physicists P.L. Dulong and A.T. Petit, states the classical expression for the molar specific heat capacity of a crystal. Experimentally the two scientists had found that the heat capacity per weight (the mass-specific heat capacity) for a number of substances became close to a constant value, after it had been multiplied by number-ratio representing the presumed relative atomic weight of the substance. These atomic weights had shortly before been suggested by Dalton. In modern terms, Dulong and Petit found that the heat capacity of a mole of many solid substances is about 3*R*, where *R* is the modern constant called the universal gas constant. Dulong and Petit were unaware of the relationship with *R*, since this constant had not yet been defined from the later kinetic theory of gases. The value of 3*R* is about 25 joules per kelvin, and Dulong and Petit essentially found that this was the heat capacity of crystals, per mole of atoms they contained. HMB theory gives theoretical base for this law (see point 3.2 of the chapter).

Thermal expansion law. To a first approximation, the change in length measurements of an object ("linear dimension" as opposed to, e.g., volumetric dimension) due to thermal expansion is related to temperature change by a "linear expansion coefficient". It is the fractional change in length per degree of temperature change. The point 4.1 presents detail modeling of this law. Heat expansion of electronic shells (including external van der Waals spheres) on linear law is described by corresponding solutions change of equation (4) for polarized atom space when changing Debye radius $D = (kT\varepsilon_0 / n_0 q^2)^{1/2} \sim T$ (since typical gaseous HMB concentration n_0 at similar pressure p_0 is inversely proportional to temperature *T*). This law works well as long as the linear-expansion coefficient does not change much over the change in temperature.

6.2. Deoxyribonucleic acid "flash" memory

Deoxyribonucleic acid (DNA) is a nucleic acid containing the genetic instructions used in the development and functioning of all known living organisms. The DNA segments carrying this genetic information are called genes. Likewise, other DNA sequences have structural purposes, or are involved in regulating the use of this genetic information. Along with ribonucleic acid and proteins, DNA is one of the three major macromolecules that are essential for all known forms of life. DNA consists of two long polymers of simple units called nucleotides, with backbones made of sugars and phosphate groups joined by ester bonds. These two strands run in opposite directions to each other and are therefore anti-parallel. Attached to each sugar is one of four types of molecules called nucleobases (informally, *bases*). It is the sequence of these four nucleobases along the backbone that encodes information. This information is read using the genetic code, which specifies the sequence of the amino acids within proteins. The code is read by copying stretches of DNA into the related nucleic acid RNA in a process called transcription. Within DNA cells is organized into long structures called chromosomes. During cell division these chromosomes are duplicated in the process of DNA replication, providing each cell its own complete set of chromosomes.

Hydrogen bonds play an important role in maintaining two circuits of DNA molecule heliciform structure (Figure 20). This distribution provides minimal summary potential energy. The principle of minimal potential energy leads to stationary localization of electrons in antiprism vertexes (the point 4.3 of the chapter). The turn of antiprism bases lead serves as a basis for forming of spiral complex molecules (in this case DNA, RNA). Common polarized space of HMB medium in DNA molecule presents a base of "flash" memory. Figure 20 shows the polarized space for *N---H---N* bond. We can propose HMB is elementary bit of information for the development and functioning of living organisms.

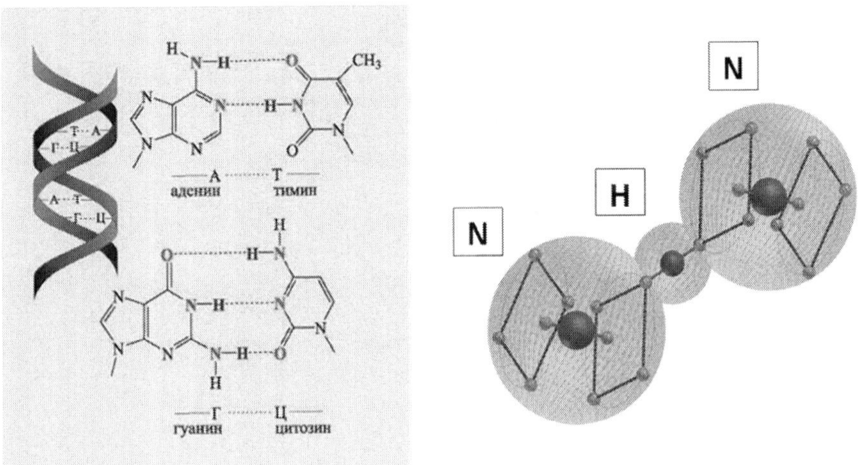

Figure 20. Hydrogen bonding N---H---N of DNA molecule

6.3. High temperature air breathing engine data

The united mathematical model (1) of the entire engine flow path is considered below (details present in [87, 88]). The typical flow passage of aviation gas turbine engine in meridional plane (z,r) is shown in figure 21. The main parts of engine (here it is a turbojet bypass engine with afterburner) are: fan (low pressure compressor), high pressure compressor (with a few variable guide vanes), main angular combustor chamber, high pressure turbine, low pressure turbine, afterburner, secondary contour, variable-area nozzle. A simulation of working fluid (air or gas) moving is fulfilled in the whole flow of engine including core and bypass duct. We use in this case equation system (1) with thin layer viscous approaches and differential turbulence models.

Results of flow simulation for bypass gas turbine engine show on figure 22-24. The engine was investigated in detail experimentally. Both the whole engine and the core engine were tested. The experiments demonstrated significant discrepancy between the tested and design engine parameters for a number of working points. In became apparent first of all in the decreased flow capacity of the compressor, and led finally to increased turbine inlet temperatures, decreasing thrust etc. Examples of compressor and turbine flows are shown on figure 23 and 24.

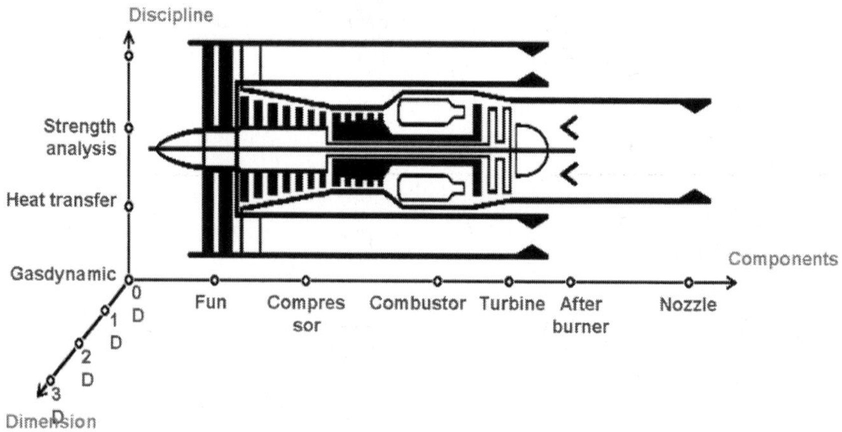

Figure 21. Air Breathing engine scheme (bypass turbojet)

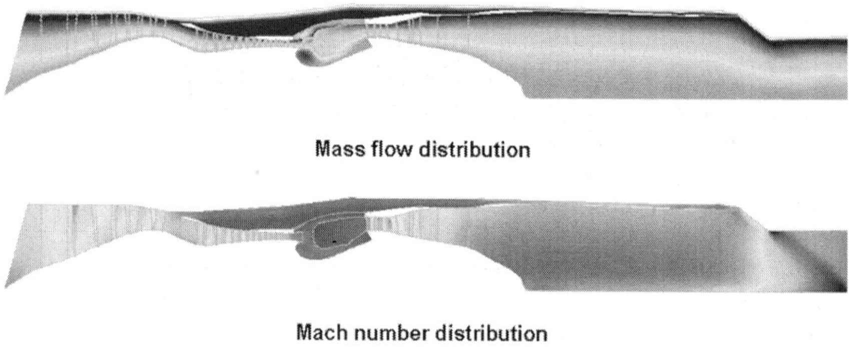

Figure 22. Examples of steady and unsteady calculations in turbojet engine

Channel flow with heat addition simulation is presented on figure 25. We can see temperature and pressure distributions inside channel and on up and down walls.

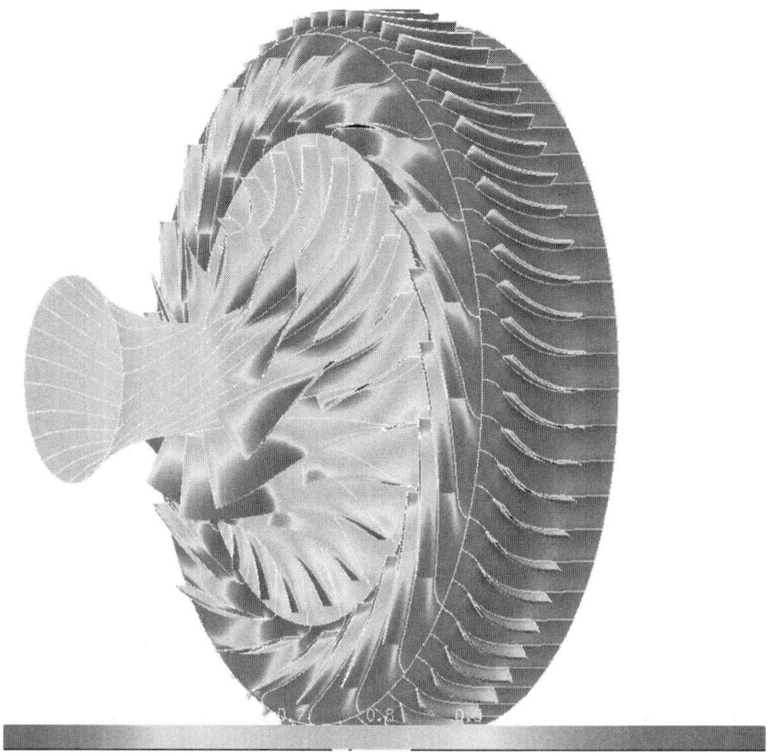

Figure 23. Static pressure distributions in the centrifugal compressor

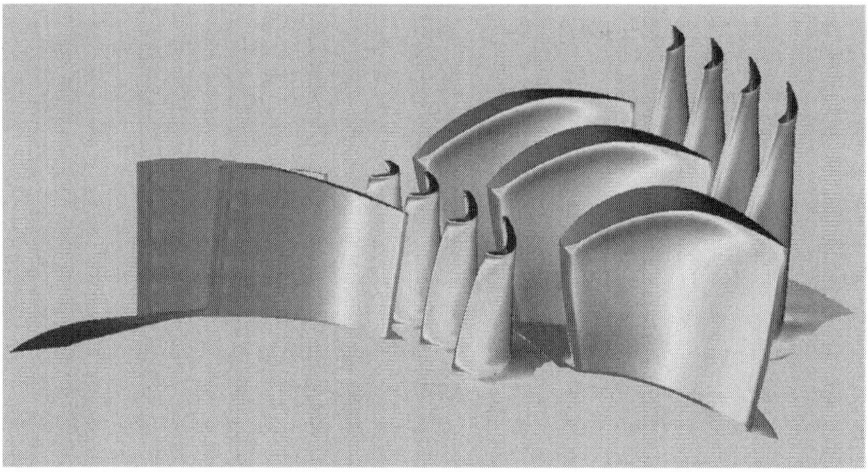

Figure 24. Mach number contours in high pressure two stage turbine

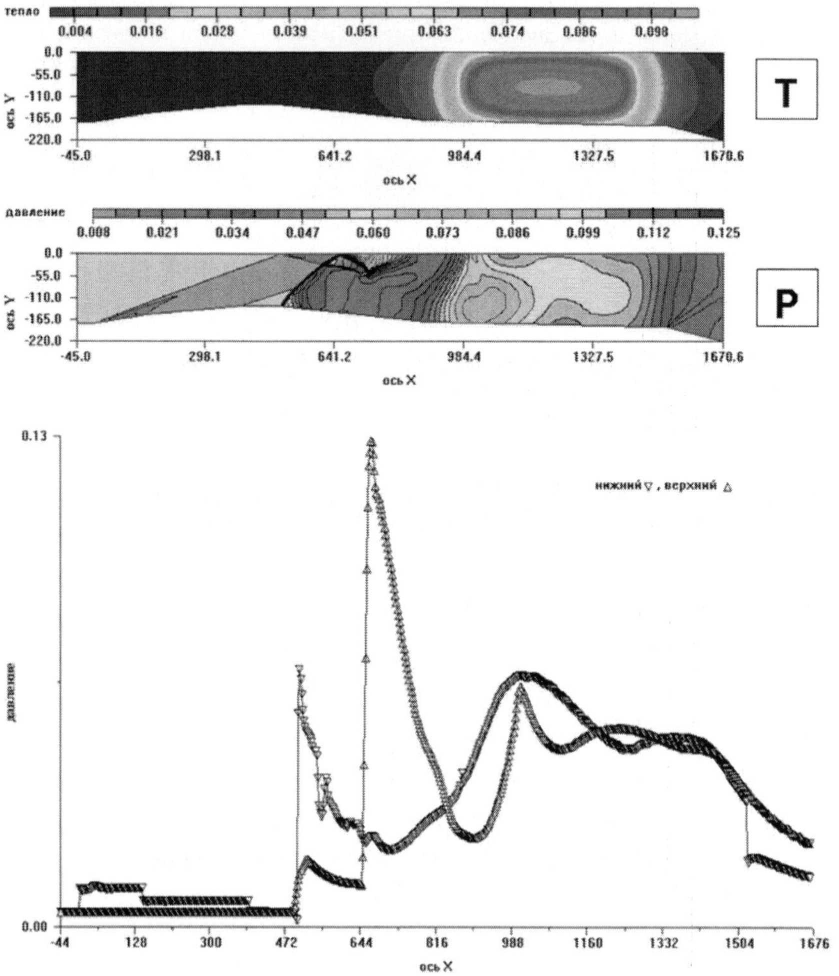

Figure 25. Pressure distributions in channel walls with intensive heat addition

6.4. Astrophysics examples

Let's demonstrate now possible application our modeling within the frame of classical mechanics approach for two astrophysics phenomena (gamma ray bursts and cosmic jets).

Gamma ray bursts. A good theoretic model of gamma-bursts can be represented by the above considered superluminal soliton solutions of rather big amplitude, which move in the dark matter medium without changing their shapes and loss of energy. We also give calculated solutions of the decomposition task of initial contraction for sequence of soliton solutions (Figure 26). As a result of such decomposition a sequence solitons with decreasing amplitude and propagation velocity is formed. The similar solutions can simulate the effects of gamma-bursts after glowing, which later reaches an observer (with the delay of several years).

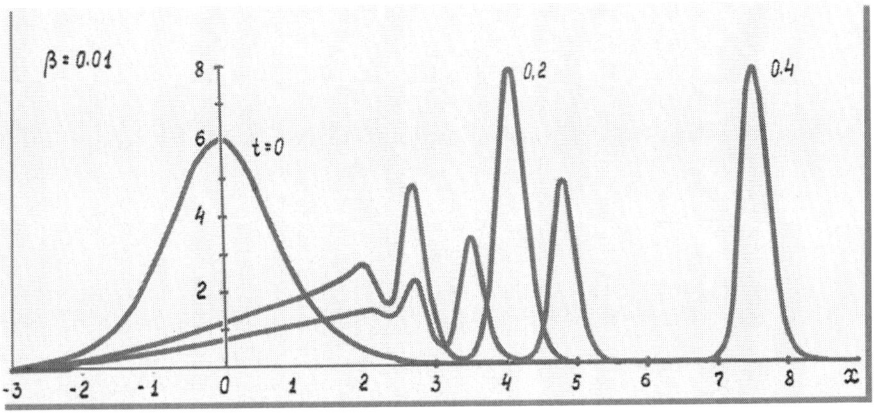

Figure 26. Gamma-ray burst and afterglow simulation

Cosmic extragalactic jets. Further using our modeling we can imagine a quasar center (of an active galaxy) like a jet engine combustion chamber form which in the born opposite directions and perpendicular to the galaxy disk the two supercritical jets are breaking out. The combustion processes of "sub baryon fuel" are going on in the quasar center at huge temperatures and pressures. Such a "sublimation" of dark holes seems more natural (than their quantum evaporation) and is well described from the standpoint of classical mechanics (obviously, an adequate definition of state equation is required). Figure 27 displays a scheme of an active galaxy with two superluminal jets flowing from its center in opposite directions. The calculation results (on equations (1)) of several "barrels" for such extragalactic jets are given (as constant pressure lines).

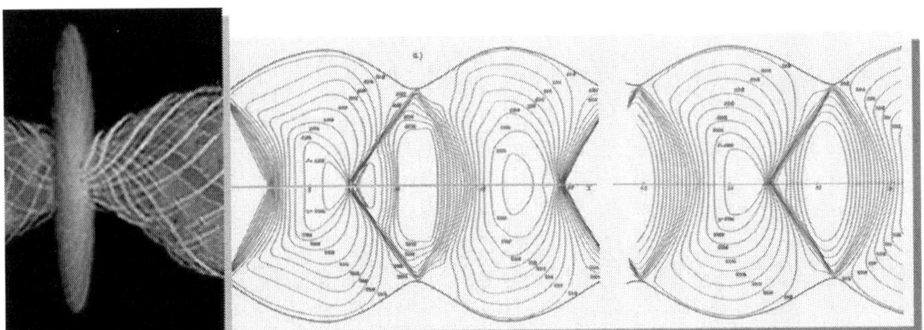

Figure 27. Natural jet engines with black hole combustor

7. Conclusion

1. Experimentally registries in the second half of XX century Cosmic Microwave Background (demonstrating the Space Energy presence) with temperature *2.725 K* and Dark Matter are the background of the Grand Unified Theory (GUT) for nature laws and phenomena. The Hidden Mass Boson in classic dipole form gives a common base for Light and Dark Matter theory.
2. The GUT uses classic Newtonian mechanics, Euler's equations of motion (conservation laws of mass, momentum and energy), Maxwell's electrodynamics and Boltzmann's kinetic theory in absolute 3D Euclidean space with absolute positive direction time.
3. The GUT on the single position describes motion matter and antimatter, electromagnetic, weak and strong interactions and electrovalence linkages (atomic and molecule structures).
4. Additional empirical confirmations of the GUT are Hooke's, Dulong–Petit's, thermal expansion laws and interaction of radiation and matter. In the plane of practical applications the corrected theory of high temperature air breathing engines is demonstrated.

Author details

Mikhail Ja. Ivanov
Gas Dynamics Department, Central Institute of Aviation Motors, Moscow, Russia

Acknowledgement

Author expresses his sincere thanks to V.I. Kokorev and B.O. Muravyov for their help in preparation of the chapter.

8. References

[1] Rubin V. Dark Matter in the Universe. Scientific American: 1998.
[2] Spooner N.J.C., Kudryavtsev V., editors. Proc. of the Third Int. Workshop on the Identification of Dark Matter. World Scientific, 2001.
[3] Dark Matter 2002. Nuclear Physics B. (Proc. Suppl.): 2003.
[4] Moskowitz G. Dark Matter hides, physicists seek. Stanford Report 2006.
[5] Mavromatos N. Recent results from indirect and direct dark matter searches – Theoretical scenarios. In: 13th ICATPP Conference. 3-7 Oct. 2011, Villa Olmo, Como, Italy.
[6] Spooner N.J.C., Kudryavtsev V., editors. Proc. of the Fourth Int. Workshop on the Identification of Dark Matter. World Scientific, 2003.
[7] Cline D.B. Sources and Detection of Dark Matter and Dark Energy in the Universe. Proceedings of the IV Int. Sym. CA, USA, Feb. 23-25, 2000.

[8] Chernin A.D. Dark energy and universe antigravitation. Uspekhi Fizicheskikh Nauk 2008; 178(3), 267-300.

[9] Riess A.G., Filippenko A.V. et al. Observational Evidence from Supernovae for an Accelerating Universe and a Cosmological Constant. The Astronomical Journal 1998; 116 (3), 1009-1038.

[10] Perlmutter S., Schmidt B.P. and Riess A.G. For the discovery of the accelerating expansion of the Universe through observations of distant supernovae. The Nobel Prize in Physics 2011.

[11] Southworth. G.C. Hyper-frequency waveguides – General considerations and experimental results. Bell Syst. Tech. Journal 1936; 15 284-309.

[12] Barrov WA.L. Transmission of electromagnetic waves in hollow tubes of metals. Proc. IRE. 1936; 24 1298-1328.

[13] Schelkunoff S.A. Transmission theory of plane electromagnetic waves. Proc. IRE, vol. 25, pp. 1457-1493, Nov. 1937.

[14] Chu L.J., Barrov W.L. Electromagnetic waves in hollow metal tubes of rectangular cross section, Proc. IRE, 1937; 26 1520-1555.

[15] Kemp J. Electromagnetic Waves in Metal Tubes of Rectangular Cross-section, Jour. I.E.E., Part III, Vol. 88, No. 3, pp. 213-218, Sept. 1941. Waveguide Transmission.

[16] Basov N.G.et al. Nonlinear amplification light pulses. Journal Exp. & Theor. Phys. 1966; 50 (1) 23-34.

[17] Krukov P.G., Letokhov V.S. Light pulse propagation in resonance amplification media. Uspekhi Fizicheskikh Nauk 1969; 99 (2)

[18] Wang L.J., Kuzmich A., Dodariu A. Gain-assisted superluminal light propagation. Nature 2000; 406 277-279.

[19] Chiao R.Y. Superluminal (but causal) propagation of wave packets in transparent media with inverted atomic populations. Phys. Rev. 1993; A 48 34-37.

[20] Yablonovitch E. Photonic band-gap crystals. Journal Phys. Condens. Matter, 1993; 5 2443.

[21] Steinberg A.M., Kwiat P.G. and Chiao R.Y. Measurement of the single-photon tunneling time. Phys. Rev. Lett., 1993; 71 708.

[22] Enders A. and Nimtz G. On superluminal barrier traversal. Journal. Phys. 1993; I2 1693-1698.

[23] Nimtz G. Evanescent modes are not necessarily Einstein causal. Eur. Phys. Journal 1999; B 7 523-525.

[24] Nimtz G., EndersA. and Spieker H. Photonic tunneling times. Journal Phys. I (France) 1994; 4 565.

[25] Alexeev I., Kim K.Y., Milchberg H.M. Phys. Rev. Lett. 2002; 88, 073901.

[26] Mugnai D., Ranfagni A. and Ruggeri R. Observation of superluminal behaviors in wave propagation. Phys. Rev. Lett. 2000; 84 4830-4833.

[27] Bigelow N.P., Hagen C.R. Comment on Observation of Superluminal Behaviors in Wave Propagation. Phys. Rev. Lett., 2001; 87 (5).

[28] Fisher D.L., Tajima T. Superluminous laser pulse in an active medium. Phys. Rev. Lett. 1993;71 4338– 4341.

[29] Oraevckiy A.N. Superluminal waves in amplification media. Uspekhi Fizicheskikh Nauk 1998; 168 (12) 1311-1321.

[30] Malakhov Ju.I., Ivanov M.Ja., Shi N.Q., Schaulov V.V. Registration of temperature dependence for electromagnetic front velocity with theoretical support and demonstration examples. Proceedings of the XI Int. Conf. (ZST-2012), RFNC-VNIITF, Apr. 16-20 2012, Snezhinsk, Russia.

[31] Mamaev V.K., Ivanov M.Ja. Electromagnetic energy flux, displacement current and polarization in physical vacuum with non zero temperature. Proceedings of the XI Int. Conf. (ZST-2012), RFNC-VNIITF, Apr. 16-20 2012, Snezhinsk, Russia.

[32] Ivanov M.Ja. Dark Matter – Quo Vadis? Proceedings of the ICATPP Conferences, 3-7 October 2011, Villa Olmo Como, Italia.

[33] Ivanov M.Ja. Classic Dark Matter Theory with Experimental Confirmations, Exact Solutions and Practical Applications. Cosmology. Proceedings of the 47-th Rencontres de Moriond, 10 – 17 March, 2012, La Thuile, Aosta valley, Italy.

[34] Higgs P.W. Broken Symmetries, Massless Particles and Gauge Fields. Physics Letters 1964; 12 132-133.

[35] 13-th ICATPP Conference on Astroparticle, Particle, Space Physics and Detectors for Physics Application. October 3-7 2011, Villa Olmo, Como, Italy.

[36] Standard Model Higgs: to be or not to be. QCD and High Energy Interactions. Proceedings of the 47-th Rencontres de Moriond. March 10 – 17, 2012, La Thuile, Aosta valley, Italy..

[37] Shipsey I. Search for Dar Matter & Higgs Boson at the LHC. Cosmology. Proceedings of the 47-th Rencontres de Moriond. March 10 – 17, 2012, La Thuile, Aosta valley, Italy.

[38] Beyond the Standard Model of Particle Physics. Rencontres du Vietnam, 15-21 July 2012, Quy Nhon Vietnam.

[39] Ivanov M.Ja., Zhestkov G.B. Dimensional Analysis, Thermodynamics and Conservation Laws in a Problem of Radiation Processes Simulation. Journal of Mathematical Research 2012; 4 (2) 10-19.

[40] Ivanov M.Ja., Mamaev V.K. Hidden mass boson. Journal Modern Physics 2012; 3(8).

[41] Politov M.V., editor. Lucretius Carus. De Rerum Nature. M.: Informconvertion, 2005.

[42] Lomonosov M.V.. Meditationes de caloris et frigoris causa. M.-L., Acad. of Sci. USSR, 1951.

[43] Boltzmann L. Lectures on gas theory. M., 1956.

[44] Maxwell J.C. A treatise on electricity and magnetism. M., Nauka, 1989.

[45] Tolman R.C. Relativity, thermodynamics and cosmology. Oxford, 1989.

[46] Landay L.D., Lifshitc E.M. The field theory. M.: Nauka, 1973.

[47] Shmaonov T.A. Methodology of Absolute Measurements for Effective Radiation Temperature with Lower Equivalent Temperature. Apparatuses and technique of experiment, 1957; 1 83-86.

[48] Penzias A.A., Wilson R.W.. A Measurement of Excess Antennatemperature at 4080 m/s. Astrophys. Journal 1965; 142 419-421.

[49] Dolgov A.D., Zeldovich Ja.B., Sagin M.V. Cosmology of Earlier Universe. Moscow: MSU, 1988.

[50] Smooth G.F. Anisotropy of Background Radiation. Uspekhi Fizicheskih Nauk 2007; 177(12) 1294-1318.

[51] Bridgmen P.W. Dimensional analysis. New Haven. Yale Univ. Press; 1932.

[52] Sedov L.I. Methods of similarity and dimension in mechanics. M., Nauka; 1967.

[53] Birkhoff G. Hydrodynamics. A study in logic, fact and similitude. Princ. Univ. Press; 1960.

[54] Buckingham E. On physically similar systems; the use of dimensional equations. Phys. Rev. 1914 4.

[55] Peccei R.D., Quinn H.R. Phys. Rev. D. 1977; 16 1971.

[56] Choi K., Kang K., Kim J.E. Phys. Lett. 1989; 62 849.

[57] Fishman. G.J. Gamma-ray bursts: an overview. Publ. Astron. Soc. Pac.; 1995.

[58] Fishman G.J., Hartmann D.H. Gamma-ray bursts. Scientific American 1997; 7.

[59] Gamma-ray bursts found to be most energetic event in universe; PRC 1998 Http://opositr.stsci.edu/publinfo/pr/1998/17/.

[60] .Amali A et al. Discovery of a transient absorption edge in the X-ray spectrum of GRB 990705. Science 2000 3 953-955.

[61] Piro L. et al. Observation of X-ray lines from a gamma-ray bursts (GRB 991216). Science 2000; 3 955-958.

[62] Piro L. The afterglow of gamma-ray bursts: Light of the mistery. http://www.ias.rm.cnr.it/ias-home/sax/cretaweb.html.

[63] Efimov N.N. et al. Proceedings of Int. Workshop on Astrophysical Aspects of the Most Energetic Cosmic Rays, Kofu 1990.

[64] Hayashida N. et al., Phys. Rev. Lett. 1994, 73, 3491.

[65] Takeda M. et al. Phys. Rev. Lett., 1998, 81, 1163.

[66] Greisen K. Phys. Rev. Lett. 1966, 2, 748.

[67] Quasars and active galaxies. Univ. of California, San Diego. http://cassfos02.uscd.edu/public/tutorial/Quasars.html.

[68] Wiita P.J. Cosmic Radio Jets. Astro-ph/0103020.

[69] Superluminal motion in compact radio sources. http://www.ira.bo.cnr. it/~tventuri/vbli2.html.

[70] VLA reveals vital details of superfast cosmic jets. http://info.aoc. nrao.edu/pr/vla20/jets.html.

[71] Quasar giant jet.10.01.2002. http://www.nature.ru/db/msg.html?mid=1177503&s.

[72] Putterman S.J. Sonoluminescence.: Sound into light. Scientific American 1995; 272 (2) 46–51. Bibcode.

[73] Chen W., Huang W., Liang Y., Gao X. and Cui W. Time-resolved spectra of single-bubble sonoluminescence in sulfuric acid with a streak camera. Phys. Rev. 2008; E 78 03530.

[74] Newton I. Papers and Letters in Natural Philosophy, ed. by I. Bernard Cohen. Harvard University Press, 1958, 1978. ISBN 0-674-46853-8.

[75] Eddington A. The Nature of the Physical World. Ann Arbor: Univ. of Michigan Press; 1958.

[76] Bernardis P. et al. A flat Universe from high-resolution maps of the cosmic microwave background radiation. Nature 2000; 404.

[77] Netterfield C.B. et al. A measurement by BOOMERANG of multiple peaks in the angular power spectrum of the cosmic microwave background. 2001. (astro-ph/0104460).

[78] Voigt W. Uber Das Dopplersche Prinzip. Gott. Nachr. 1887; 41.

[79] Ivanov M.Ja. To analogy of gas dynamics and electrodynamics models. Journal Fizicheskaya Misl Rossii 1998; 1 1-14.

[80] Ivanov. M.Ja. Dynamics of vector force fields in a free space. RAS: Mathematical simulation 1998; 10 (7) 3-20.

[81] Burke D.L. et al. Positron production in multiphoton light-by-light scattering.Phys. Rev. Let. 1997; 79(1), 1626–1629.

[82] Isihara A. Statistical Physics. Acad. Press: NY-L, 1973.

[83] Anselm A.I. Bases of statistical physics and thermodynamics. M.: Nauka; 1973.

[84] Ivanov M.Ja. Thermodynamically compatible conservation laws in the model of heat conduction radiating gas. Comp. Math. and Math. Phys. 2011; 51(1) 133-142.

[85] Marble F. Dynamics of dusty gases. Ann. Rev. of Fluid Mech. 1970; 2 397-445.

[86] Loytcansky L.G.. Mechanics of fluid and gas. M.: Nauka, 1973.

[87] Ivanov M.Ja.,Nigmatullin R.Z.. Simulation of Working Processes in Gas Turbine Engine Passage . Successes Mechanics, Vladivostok, 2009..

[88] Ivanov M.Ja, Mamaev B.I., Nigmatullin R.Z..United Modeling of Working Process in Aircraft Gas Turbine Engines , ASME Paper 2008; 50185 10 p.

[89] Okun L.B. Physics of elementary particles M.: URSS, 2008.

[90] Ivanov M.Ja. Physical models of van der Waals atomic spheres and molecule structures. M.: Conversion in Machine Building of Russia 2008; 2(87) 35-41.

[91] Ivanov M.Ja., Malinin A.V., Yanovskiy L.S. On development of electronic theory of valence linkages. M.: Ramjets & Chemmotology. CIAM Proceeding No.1340, 2010.

[92] Ivanov M.Ja., Terentieva L.V. Elements on gas dynamics of dispersion medium. M.: Informconvertion; 2004.

[93] Ivanov M.Ja., Terentieva L.V. Exact solutions of two-fluid approach equations in aerospace plasmadynamics. AIAA Paper No. 2003-0843, 8.

[94] Ivanov M.Ja., Terentieva L.V. Soliton –like structures in Dark Matter. Nuclear Physics B, 2003; 124 148-151.

[95] Ivanov M.Ja., Terentieva L.V. Particle-wave aspects of Dark Matter.Exploring the Universe. Proc. XXXIXth Ren. de Moriond, 2004.

Chapter 2

Barotropic and Baroclinic Tidal Energy

Dujuan Kang

Additional information is available at the end of the chapter

http://dx.doi.org/10.5772/52293

1. Introduction

Oceans cover approximately 71% percent of the Earth's surface, which makes them the largest solar energy collector in the world. On the other hand, the seawater moves constantly at different scales ranging from large-scale ocean currents down to centimeter-scale turbulent motions. These movements create a huge store of kinetic energy in the ocean. The ocean energy is clean and renewable. Therefore, a better understanding of the physical processes that govern the ocean movements is crucial to utilize the ocean energy more efficiently.

The tides are one of the major sources of energy to mix the interior ocean. The barotropic tidal energy is converted into heat through a series of important mixing processes. When the barotropic tides flow over rough topographic features, a portion of the barotropic energy is lost directly through local mixing, while the other portion is converted into baroclinic energy through the generation of internal (baroclinic) tides. This generated baroclinic energy either dissipates locally or radiates into the open ocean, and then cascades into smaller scales along the internal wave spectrum and finally turns into deep ocean turbulence (Figure 1). In the past decade significant efforts have been made to understand these tidal mixing processes and the associated energy distributions. Munk & Wunsch (1998) proposed a global tidal energy flux budget as shown in Figure 1. Of the 3.5 TW (1 TW = 10^{12} W) total tidal energy lost in the ocean, approximately 2.6 TW is dissipated in shallow marginal seas through bottom friction, while the remaining portion is lost in the deep ocean. Egbert & Ray (2000, 2001) have confirmed that approximately 1 TW, or 25–30% of the global total tidal energy is lost in the deep ocean by inferring dissipation from a global tidal model. They found the tides lose much more energy in the open ocean, generally in regions with rough topographic features (Figure 2). Field observations also showed that turbulent mixing is several order of magnitude larger over rough topography than over smooth abyssal plains [18]. This evidence has led to the interest in internal tides as a major source of energy for deep-ocean mixing. Both analytical and numerical investigations have also been performed to estimate the tidal energy flux budget [1, 9, 17, 20].

This chapter focuses on the numerical investigation of tidal energy in the ocean. The problem of how and where the ocean tides distribute their energy is discussed using a numerical study

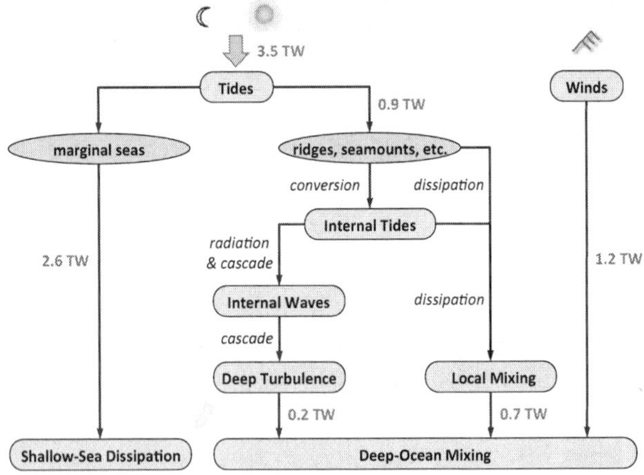

Figure 1. Global energy flux budget based on Munk and Wunsch (1998). The tides and winds are the two major sources of energy to mix the ocean. The tides contribute 3.5 TW of energy with 2.6 TW dissipated in shallow marginal seas and 0.9 TW lost in the deep ocean. The winds provide 1.2 TW of additional mixing power to maintain the global abyssal density distribution.

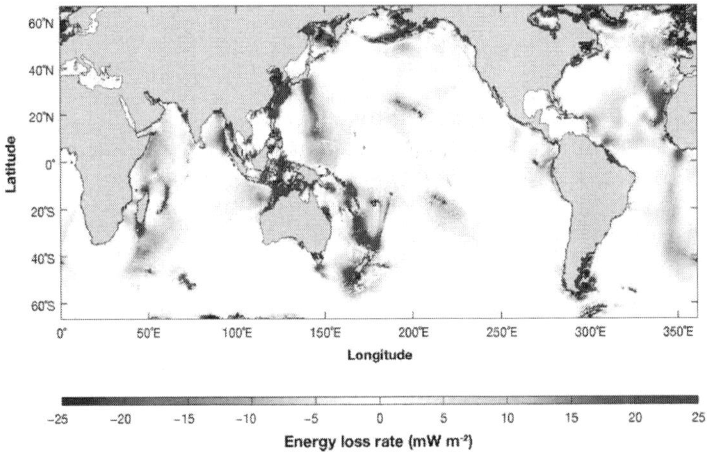

Figure 2. Global tidal energy loss rate derived from satellite altimetry sea-surface elevation data (Egbert and Ray, 2001).

of the tidal energetics in the Monterey Bay area along the central west coast of the United States. The purpose of this work is twofold: first, to provide a theoretical framework for the accurate evaluation of tidal energy budget and, second, to conduct numerical simulations of barotropic and baroclinic tides for a given region in the ocean and estimate the tidal energy budget based on the theoretical framework. A brief derivation of the barotropic and baroclinic energy equations is presented in Section 2. Subsequent sections focus on the numerical simulations which include the model setup and tidal dynamics in Section 3, and the energetics analysis in Section 4. The characteristics of tidal energy conversion is examined in Section 5. Finally, conclusions are summarized in Section 6.

2. Theoretical framework

In order to study the energetics of barotropic and baroclinic tides, we derive the barotropic and baroclinic energy equations as a theoretical framework for the numerical evaluation of the tidal energy budget in subsequent sections. Here we provide a brief description of the equations. More detailed derivation can be found in Kang (2010).

The derivation is based on the three-dimensional Reynolds-averaged Navier-Stokes equations under the Boussinesq approximation, along with the density transport equation and the continuity equation,

$$\frac{\partial \mathbf{u}}{\partial t} + \mathbf{u} \cdot \nabla \mathbf{u} = -2\boldsymbol{\Omega} \times \mathbf{u} - \frac{1}{\rho_0}\nabla p - \frac{g}{\rho_0}\rho\mathbf{k} + \nabla_H \cdot (\nu_H \nabla_H \mathbf{u}) + \frac{\partial}{\partial z}\left(\nu_V \frac{\partial \mathbf{u}}{\partial z}\right), \quad (1)$$

$$\frac{\partial \rho}{\partial t} + \mathbf{u} \cdot \nabla \rho = \nabla_H \cdot (\kappa_H \nabla_H \rho) + \frac{\partial}{\partial z}\left(\kappa_V \frac{\partial \rho}{\partial z}\right), \quad (2)$$

$$\nabla \cdot \mathbf{u} = 0, \quad (3)$$

where $\mathbf{u} = (u, v, w)$ is the velocity vector and $\boldsymbol{\Omega}$ is the Earth's angular velocity vector. ν and κ, in units of $m^2\ s^{-1}$, are the eddy viscosity and eddy diffusivity, respectively. $(\ )_H$ and $(\ )_V$ are the horizontal and vertical components of a variable or operator. The total density is given by $\rho = \rho_0 + \rho_b + \rho'$, where ρ_0 is the constant reference density, ρ_b is the background density, and ρ' is the perturbation density due to wave motions. The pressure is split into its hydrostatic (p_h) and nonhydrostatic (q) parts with $p = p_h + q$, where the hydrostatic pressure can be further decomposed with $p_h = p_0 + p_b + p'$. To obtain the barotropic and baroclinic, the velocity is also split into its barotropic and baroclinic parts as $\mathbf{u} = \mathbf{U} + \mathbf{u}'$. Accordingly, the kinetic energy density, in units of $J\ m^{-3}$, is decomposed as

$$E_k = \underbrace{\frac{1}{2}\rho_0\left(U^2 + V^2\right)}_{E_{k0}} + \underbrace{\frac{1}{2}\rho_0\left(u'^2 + v'^2 + w^2\right)}_{E'_k} + \underbrace{\rho_0\left(Uu' + Vv'\right)}_{E'_{k0}}, \quad (4)$$

where E_{k0} is the barotropic horizontal kinetic energy density, E'_k is the baroclinic kinetic energy density, and E'_{k0} is the cross term which vanishes upon depth-integration. Following Gill (1982), the perturbation potential energy due to surface elevation, in units of $J\ m^{-2}$, is given by

$$\overline{E_{p0}} = \frac{1}{2}\rho_0 g \eta^2, \quad (5)$$

in which η is the free surface elevation. The available potential energy density, in units of $J\ m^{-3}$, is defined as

$$E'_p = \int_{z-\zeta}^{z} g\left[\rho_b(z) + \rho'(z) - \rho_b(z')\right] dz', \quad (6)$$

where ζ is the vertical displacement of a fluid particle due to wave motions. This definition is an exact expression of the local APE because it computes the true active potential energy between the perturbed and unperturbed density profiles [12, 14].

Applying the variable decompositions and the boundary conditions, we obtain the depth-integrated barotropic and baroclinic energy equations as

$$\frac{\partial}{\partial t}\left(\overline{E_{k0}} + \overline{E_{p0}}\right) + \nabla_H \cdot \overline{\mathbf{F}_0} = -\overline{C} - \overline{\epsilon_0} - D_0\,, \tag{7}$$

$$\frac{\partial}{\partial t}\left(\overline{E'_k} + \overline{E'_p}\right) + \nabla_H \cdot \overline{\mathbf{F}'} = \overline{C} - \overline{\epsilon'} - D'\,, \tag{8}$$

where the depth-integrated barotropic and baroclinic energy flux terms, with the small unclosed terms neglected, are given by

$$\overline{\mathbf{F}_0} = \underbrace{\mathbf{U}_H\overline{E_{k0}}}_{\text{Advection}} \underbrace{+\mathbf{U}_H H\rho_0 g\eta + \mathbf{U}_H\overline{p'} + \mathbf{U}_H\overline{q}}_{\text{Pressure work}} \underbrace{-\overline{\nu_H \nabla_H E_{k0}}}_{\text{Diffusion}}\,, \tag{9}$$

$$\overline{\mathbf{F}'} = \underbrace{\overline{\mathbf{u}_H E'_k} + \overline{\mathbf{u}_H E'_{k0}} + \overline{\mathbf{u}_H E'_p}}_{\text{Advection}} \underbrace{+\overline{\mathbf{u}'_H p'} + \overline{\mathbf{u}'_H q}}_{\text{Pressure work}} \underbrace{-\overline{\nu_H \nabla_H E'_k} - \overline{\kappa_H \nabla_H E'_p}}_{\text{Diffusion}}\,, \tag{10}$$

in which the contributions from energy advection, pressure work, and diffusion have been labeled. The definitions of the barotropic-to-baroclinic conversion rate ($\overline{C}$), the dissipation rates (ϵ_0 and ϵ') and the bottom drag terms (D_0 and D') are not listed here. Please refer to Kang (2010) for details.

The time-averaged forms of equations (7) and (8) are given by

$$\frac{1}{T}\Delta\overline{E_0} + \nabla_H \cdot \left\langle\overline{\mathbf{F}_0}\right\rangle = -\left\langle\overline{C}\right\rangle - \left\langle\overline{\epsilon_0} + D_0\right\rangle\,, \tag{11}$$

$$\frac{1}{T}\Delta\overline{E'} + \nabla_H \cdot \left\langle\overline{\mathbf{F}'}\right\rangle = \left\langle\overline{C}\right\rangle - \left\langle\overline{\epsilon'} + D'\right\rangle\,, \tag{12}$$

where $\langle\cdot\rangle = \frac{1}{T}\int_t^{t+T}(\cdot)\,d\tau$ is the time-average of a quantity over a time interval T. For a periodic system with period T, $\Delta\overline{E_0}$ and $\Delta\overline{E'}$ tend to zero and thus the first term in equations (11)-(12) vanishes. The remaining terms describe the energy balance associated with tidal dissipation processes. The $\nabla_H \cdot \left\langle\overline{\mathbf{F}_0}\right\rangle$ term represents the total barotropic energy that is available for conversion to baroclinic energy, $\left\langle\overline{C}\right\rangle$ represents the portion of the barotropic energy that is converted into baroclinic energy, and the $\nabla_H \cdot \left\langle\overline{\mathbf{F}'}\right\rangle$ term represents the portion of the converted baroclinic energy that radiates from the conversion site. Local dissipation occurs along with the conversion and radiation processes, and they are measured by the barotopic ($-\left\langle\overline{\epsilon_0} + D_0\right\rangle$) and baroclinic ($-\left\langle\overline{\epsilon'} + D'\right\rangle$) dissipation terms, respectively. This approach presents an exact measure of the barotropic-to-baroclinic tidal energy conversion.

3. Numerical simulations

We can implement the theoretical framework into a numerical ocean model and then use it to analyze the tidal energy budget. In this section, we provide an example of the numerical investigation of the tidal energetics in the Monterey Bay area along the central west coast of the United States.

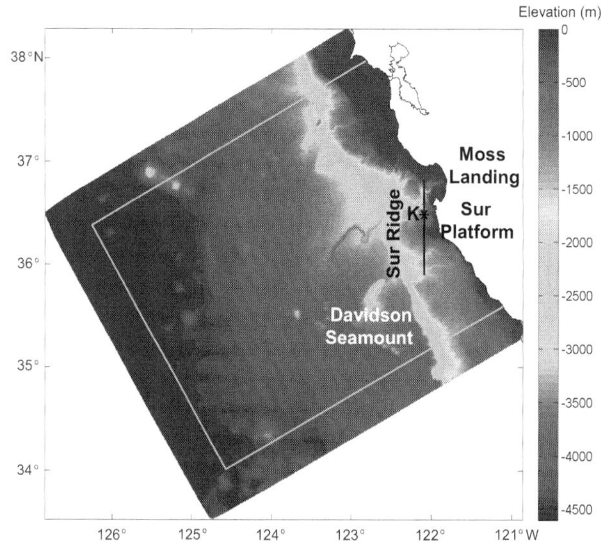

Figure 3. Bathymetry map of Monterey Bay and the surrounding open ocean. The black ∗ indicates a field observation station K. The domain outside the white box indicates the area affected by the sponge layers in the simulations. The solid black line indicates the vertical transaction along which baroclinic velocities are shown in Figure 5.

3.1. Model setup

Monterey Bay is featured by the prominent Monterey Submarine Canyon (MSC), numerous ridges, smaller canyons, and a continental slope and break region. This area is exposed to the large- and meso-scale variations of the California Current System as well as the tidal currents. Energetic internal wave activity has been observed in MSC and the surrounding region. Due to the complex bathymetry, tidal mixing processes in this area are of great interest. Our simulation domain extends approximately 200 km north and south of Moss Landing, and 400 km offshore (Figure 3), which is large enough to allow the evolution of offshore-propagating waves.

The ocean model we employ for this study is the SUNTANS model of Fringer *et al.* (2006). The resolution of the horizontal unstructured grid smoothly transitions from roughly 80 m within the Bay to 11 km along the offshore boundary. In the vertical, there are 120 z-levels with thickness stretching from roughly 6.6 m at the surface to 124 m in the deepest location, which provides better vertical resolution in the shallow regions. In total, the mesh consists of approximately 6 million grid cells in 3D.

The initial free-surface and velocity field are initialized as quiescent throughout the domain. The initial stratification is specified with horizontally-homogeneous temperature and salinity profiles. At the coastline, we apply the no-flow condition, while at the three open boundaries the barotropic velocities are specified with the OTIS global tidal model [2]. A sponge layer is imposed at each of the open boundaries to absorb the internal waves and minimize the reflection of baroclinic energy into the domain. More model setup details and the validation of the model skill can be found in Kang and Fringer (2012).

3.2. Tidal dynamics

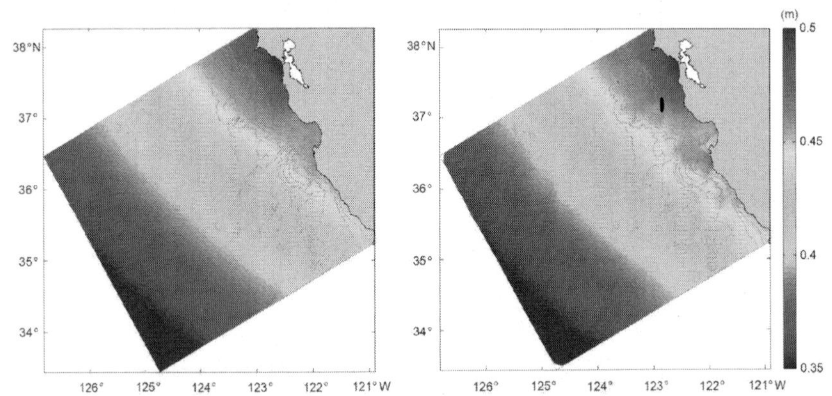

Figure 4. Sea surface elevation at $t = 18T_{M_2}$ from M_2 forced barotropic run (left) and baroclinic run (right). Bathymetry contours are spaced at -200, -500, -1000, -1500, -2000, -2500, -3000, -3500 m.

Monterey Bay is exposed to predominantly semidiurnal M_2 tide. Therefore we carry out two sets of M_2 forced simulations, one with homogeneous density to investigate the barotropic tides only, while the other with vertical stratification to study the baroclinic tides. Each simulation is run for 18 M_2 tidal cycles. Sea-surface elevations from such two runs are illustrated in Figure 4. This comparison highlights the modulation of baroclinic tides to the barotropic flow field, particularly near rough bottom topography.

Figure 5 demonstrates the vertical distribution of baroclinic velocities along a transect indicated in Figure 3. It can be seen clearly that baroclinic tides are generated at Sur Platform and then radiate to the north and south in the form of tidal beams.

4. Tidal energetics

We evaluate the depth-integrated, time-averaged barotropic and baroclinic energy equations (11) and (12) for the energy analysis in this section. They are averaged over the last six M_2 tidal cycles of the 18-M_2-cycle baroclinic simulation. Because the system is periodic, the first term in equation (11)-(12) tends to zero upon period-averaging. We therefore obtain the balance relations

$$\nabla_H \cdot \left\langle \overline{\mathbf{F}_0} \right\rangle = -\left\langle \overline{C} \right\rangle - \left\langle \overline{\epsilon_0} + D_0 \right\rangle , \tag{13}$$

$$\nabla_H \cdot \left\langle \overline{\mathbf{F}'} \right\rangle = \left\langle \overline{C} \right\rangle - \left\langle \overline{\epsilon'} + D' \right\rangle . \tag{14}$$

The model computes all the energy terms in the barotropic and baroclinic equations. In the following analysis, the conversion rate, $\langle \overline{C} \rangle$, and the energy flux divergence terms, $\nabla_H \cdot \langle \overline{\mathbf{F}_0} \rangle$

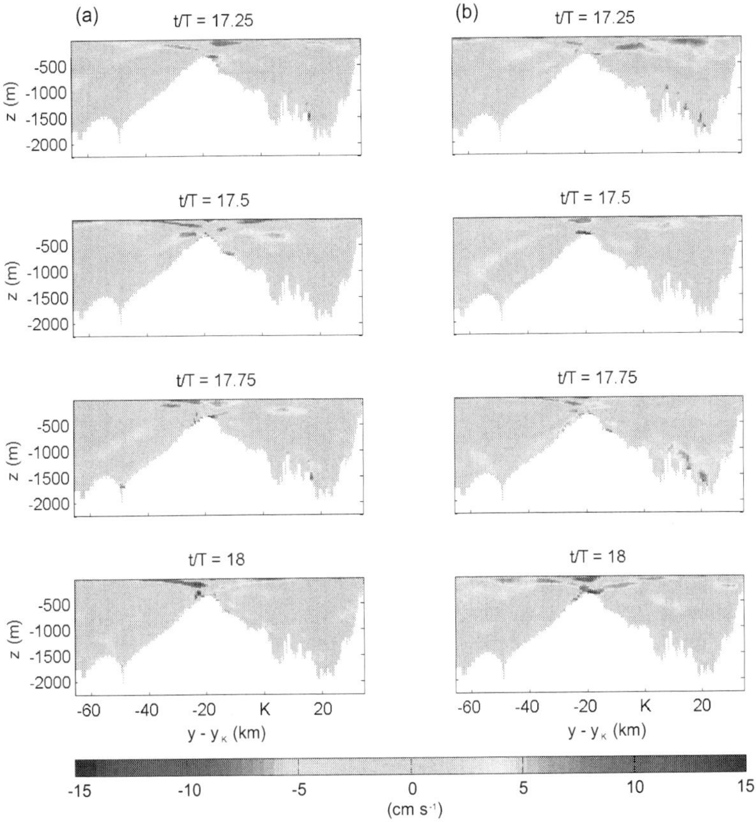

Figure 5. Vertical structure of the East-West (a) and North-South (b) baroclinic velocities along the transect indicated in Figure 3.

and $\nabla_H \cdot \left\langle \overline{\mathbf{F}'} \right\rangle$, are from direct model outputs. However, the barotropic and baroclinic dissipation rates are inferred from the above balance relations as in Niwa and Hibiya (2004).

4.1. Horizontal structure

The left panel of Figure 6 illustrates the horizontal distribution of the depth-integrated baroclinic energy flux vectors, $\left\langle \overline{\mathbf{F}'} \right\rangle$. Large energy fluxes are seen in the vicinity of rough topographical features, such as the MSC, the Sur Ridge-Platform region, and the Davidson Seamount. The right panel of Figure 6 shows the horizontal distribution of the depth-integrated barotropic-to-baroclinic conversion rate, $\langle \overline{C} \rangle$. Red color represents positive energy conversion rate, which implies generation of internal tides, and negative energy conversion rate (blue color) represents energy transfer from the baroclinic tide to the barotropic tide. Negative energy conversion is due to the phase difference between locally-

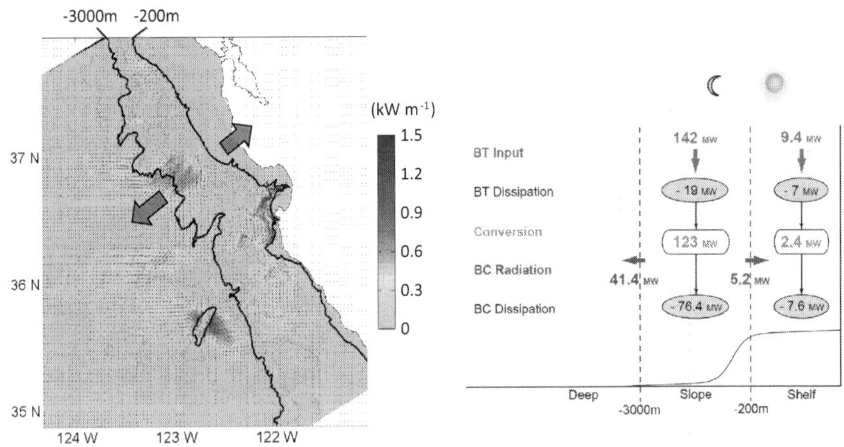

Figure 8. Schematic of the M_2 tidal energy budget for the two subdomains bounded by the 0-m, 200-m and 3000-m isobaths.

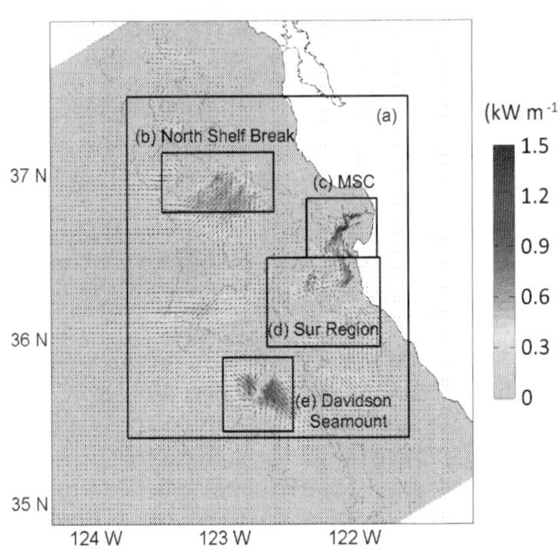

Region	η_c	η_r
(a) All	88%	42%
(b) Shelf Break	94%	76%
(c) Submarine Canyon	71%	-62%
(d) Ridge Platform	88%	61%
(e) Seamount	95%	81%

Figure 9. Efficiency of the M_2 baroclinic energy conversion and radiation for the five subdomains indicated in the left figure.

4.3. Energy flux contributions

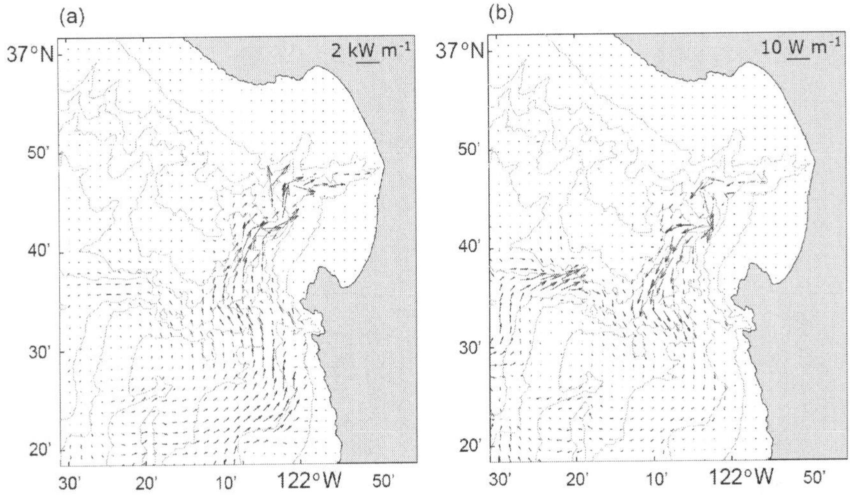

Figure 10. Baroclinic energy flux contributions from (a) hydrostatic, and (b) nonhydrostatic pressure work in the Monterey Submarine Cayon region.

As discussed in Section 2, our method computes the full energy fluxes and thus allows us to compare the contributions of different components. Here we choose subdomain (a) as our study domain. We found that the component due to hydrostatic pressure work (the traditional energy flux) is the dominant term. If we consider the total energy radiation as 100%, the hydrostatic contribution is $\sim$ 101% while the other terms account for the remaining -1%. The advection and nonhydrostatic contributions are quite small, which implies that the internal tides in the Monterey Bay area are mainly linear and hydrostatic. Figure 10 shows that the hydrostatic and nonhydrostatic energy fluxes oppose one another within MSC. This occurs because the effect of the nonhydrostatic pressure is to restrict the acceleration owing to the impact of vertical inertia. Hydrostatic models therefore tend to overpredict the energy flux particularly for strongly nonhydrostatic flows.

5. Mechanism of energy conversion

Several nondimensional parameters are generally employed to discuss the character of barotropic-to-braoclinic energy conversion. The first parameter is the steepness parameter defined by

$$\epsilon_1 = \frac{\gamma}{s}, \tag{22}$$

where γ is the topographic slope, and s is the internal wave characteristic slope. The steepness parameter is used to distinguish between subcritical ($\epsilon_1 < 1$) and supercritical topography ($\epsilon_1 > 1$). The topography is referred to as critical when $\epsilon_1 = 1$.

The second nondimensional parameter is the tidal excursion parameter defined by

$$\epsilon_2 = \frac{U_0 k_b}{\omega}, \tag{23}$$

which measures the ratio of the tidal excursion U_0/ω to the horizontal scale of the topography k_b^{-1}.The excursion parameter is used to examine the nonlinearity of the waves [6, 15, 20]. When $\epsilon_2 \ll 1$, linear internal tides are generated mainly at the forcing frequency ω. Over subcritical topography ($\epsilon_1 < 1$) most of the energy generation is in the first mode internal tide, while over critical or supercritical topography ($\epsilon_1 \geq 1$), higher modes are generated and their superposition creates internal tidal beams. At intermediate excursion ($\epsilon_2 \sim 1$), nonlinearity becomes important, and nonlinear internal wave bores, weak unsteady lee waves, and solitary internal waves may be generated depending on the topographic features. When $\epsilon_2 > 1$, in addition to bores and solitary internal waves, strong unsteady lee waves may form [22].

Although internal wave generation is a complex process, we can summarize the behavior of the internal wave generation in Monterey Bay by plotting histograms of the conversion and divergence terms as functions of the criticality and excursion parameters. Here we compute the two parameters throughout subdomain (a). The upper and left panels of Figures 11 demonstrate the distribution of conversion as a function of the nondimensional parameters ϵ_1 and ϵ_2, respectively. The energy conversion (green bins) occurs predominantly in regions within which $\epsilon_1 < 5$ and $\epsilon_2 < 0.02$. Under these conditions, baroclinic tides generated in this region are mainly linear and in the form of internal tidal beams [6, 20, 22]. As expected, conversion of barotropic energy into baroclinic energy peaks for critical topography near $\epsilon_1 \sim 1$. More interesting, however, is that there is also a peak in conversion for a particular value of $\epsilon_2 \sim 0.005$.

The lower right panel of Figure 11 depicts the distribution of the energy conversion as a function of the two parameters. Energy conversion occurs mainly in regions where ϵ_1 and ϵ_2 satisfy a particular relation. When both parameters are small, this relation is linear. As the values of the two parameters increase, the departure is weakly quadratic.

6. Summary

The tides are one of the main power to mix the ocean, which is the largest energy collector from the Earth-Moon system. A better understanding of how and where the tides lost their energy is very important to the climate and energy study. In this chapter, we present a numerical method to investigate the tidal energy conversion and estimate the tidal energy budget.

A theoretical framework for analyzing tidal energetics is derived based on the complete form of the barotropic and baroclinic energy equations that provide an accurate and detailed energy analysis. Three-dimensional, high-resolution simulations of the tides and waves in the Monterey Bay area are conducted using the hydrodynamic coastal SUNTANS ocean model. Based on the theoretical approach, model results are analyzed to address the question of how the barotropic tidal energy is partitioned between local barotropic dissipation and local generation of baroclinic energy, and then how much of this generated baroclinic energy is lost locally versus how much is radiated away for open-ocean mixing. Subdomain (a), a 200 km $\times$ 230 km domain, is used to represent the Monterey Bay area because it includes

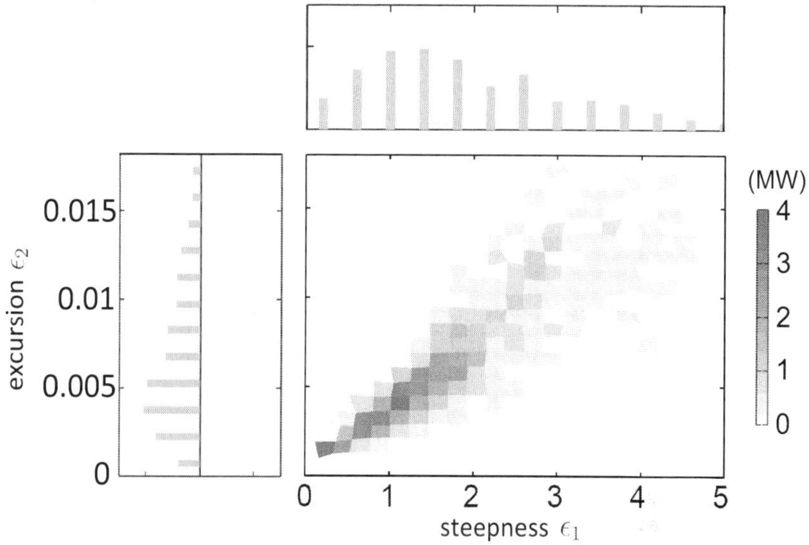

Figure 11. Distribution of the barotropic-to-baroclinic energy conversion as a function of steepness parameter ϵ_1 and excursion parameter ϵ_2 for subdomain (a).

all typical topographic features in this region. Of the 152 MW energy lost from the barotropic tide, approximately 133 MW (88%) is converted into baroclinic energy through internal tide generation, and 42% (56 MW) of this baroclinic energy radiates away for open-ocean mixing (Figure 12). The tidal energy partitioning depends greatly on the topographic features. The Davidson Seamount and the Northern shelf-break region are most efficient at baroclinic energy generation and radiation. The Sur Platform region converts a large portion and radiates roughly half of the barotropic energy as baroclinic energy. The Monterey Submarine Canyon acts as an energy sink because it does not radiate energy but instead absorbs the baroclinic energy from the Sur Platform region. The energy flux contributions from nonlinear and nonhydrostatic effects are also examined. The small advection and nonhydrostatic contributions imply that the internal tides in the Monterey Bay area are predominantly linear and hydrostatic.

We also investigate the character of tidal energy conversion by examining the energy distribution as a function of two nondimensional parameters, namely the steepness parameter ($\epsilon_1 = \gamma/s$) and the excursion parameter ($\epsilon_2 = U_0 k_b/\omega$). The generation mainly occurs in the regions satisfying $\epsilon_1 < 5$ and $\epsilon_2 < 0.02$, indicating that baroclinic tides generated in the Monterey Bay area are mainly linear and in the form of internal tidal beams. The results highlight how description of the conversion process with simple nondimensional parameters produces results that are consistent with theory, in that internal wave energy generation peaks at critical topography ($\epsilon_1 \sim 1$). The results also indicate that conversion peaks for a particular excursion parameter ($\epsilon_2 \sim 0.005$ for this case). This implies that it may be possible to parameterize conversion of barotropic to baroclinic energy in barotropic models

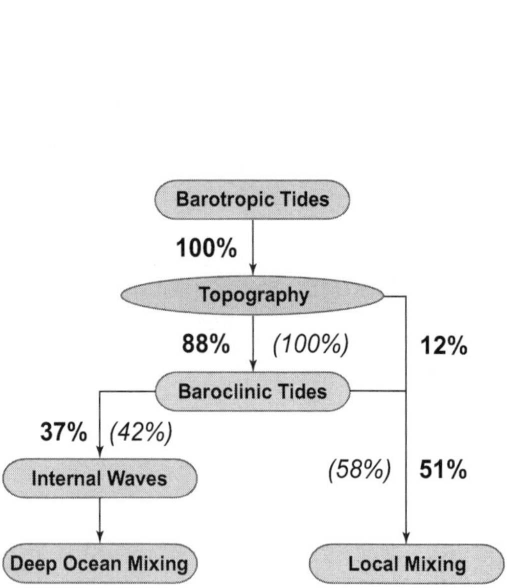

Figure 12. Schematic of the M_2 tidal energy budget in percentages for subdomain (a) indicated in Figure 9. The bold percentages are relative to the total input barotropic energy, and the thin italic percentages are relative to the generated baroclinic energy.

with knowledge of ϵ_1 and ϵ_2. For example, a parameterization of internal wave generation based on the steepness parameter has been widely used in global barotropic tidal models [21] and ocean general circulation models [10].

This work outlines a systematic approach to analyze internal tide energetics and estimate tidal energy budget regionally and globally. The results draw a picture of how the M_2 tidal energy is distributed in the Monterey Bay region. The Monterey Bay area is exposed to the large-scale California Current System and meso-scale eddies and upwelling. The seasonally varying dynamics may affect the stratification and thus the generation and propagation of internal tides in this area. Therefore, it may be necessary to consider seasonal effects of stratification and to include mesoscale effects by coupling with a larger-scale regional model such as ROMS [8, 19].

Acknowledgements

The author gratefully acknowledge the support of Professor Oliver Fringer at Stanford University and the ONR Grant. I would like to thank Samantha Terker, Drs. Jody Klymak, Robert Pinkel, James Girton, and Eric Kunze for kindly providing the field data. The helpful discussions with Drs. Robert Street, Stephen Monismith, Leif Thomas, and Rocky Geyer are greatly appreciated. I also thank Dr. Steven Jachec for useful help with simulation setup.

Author details

Dujuan Kang

Institute of Marine and Coastal Sciences, Rutgers University, New Jersey, USA

7. References

[1] Carter G.S (2010) Barotropic and baroclinic M2 tides in the Monterey Bay region. J. Phys. Oceanogr. 40: 1766-1783.

[2] Egbert G.D, Erofeeva S.Y (2002) Efficient inverse modeling of barotropic ocean tides. J. Atmos. Oceanic Technol. 19: 183-204.

[3] Egbert G.D, Ray R.D (2000) Significant dissipation of tidal energy in the deep ocean inferred from satellite altimeter data. Nature. 405: 775-778.

[4] Egbert G.D, Ray R.D (2001) Estimates of M2 tidal energy dissipation from TOPEX/Poseidon altimeter data. J. Geophys. Res. 106(C10): 22475-22502.

[5] Fringer O.B, Gerritsen M, Street R.L (2006) An unstructured-grid, finite-volume, nonhydrostatic, parallel coastal ocean simulator. Ocean Modelling. 14: 139-173.

[6] Garrett C, Kunze E (2007) Internal tide generation in the deep ocean. Annu. Rev. Fluid Mech. 39: 57-87.

[7] Gill A.E (1982) Atmosphere-Ocean Dynamics. Academic Press.

[8] Haidvogel D.B, Arango H.G, Hedstrom K, Beckmann A, Malanotte-Rizzoli P, Shchepetkin A.F, 2000: Model evaluation experiments in the North Atlantic Basin: Simulations in nonlinear terrain-following coordinates. Dyn. Atmos. Oceans. 32: 239-281.

[9] Jachec S.M, Fringer O.B, Gerritsen M.G, Street R.L (2006) Numerical simulation of internal tides and the resulting energetics within Monterey Bay and the surrounding area. Geophys. Res. Lett. 33: L12605, doi:10.1029/2006GL026314.

[10] Jayne S.R (2009) The impact of abyssal mixing parameterizations in an ocean general circulation model. J. Phys. Oceanogr. 39: 1756-1775.

[11] Kang D. (2010) Energetics and dynamics of internal tides in Monterey Bay using numerical simulations. Ph.D. Dissertation. Stanford University. 170 p.

[12] Kang D, Fringer O.B (2010) On the calculation of available potential energy in internal wave fields. J. Phys. Oceanogr. 40: 2539-2545.

[13] Kang D, Fringer O.B (2012) Energetics of barotropic and baroclinic tides in the Monterey Bay area. J. Phys. Oceanogr. 42: 272-290.

[14] Lamb K.G (2007) Energy and pseudoenergy flux in the internal wave field generated by tidal flow over topography. Cont. Shelf Res. 27: 1208-1232.

[15] Legg S, Huijts K.M.H (2006) Preliminary simulations of internal waves and mixing generated by finite amplitude tidal flow over isolated topography. Deep-Sea Res. II. 53: 140-156.

[16] Munk W, Wunsch C (1998) Abyssal recipes II: Energetics of tidal and wind mixing. Deep-Sea Res. 45: 1977-2010.

[17] Niwa Y, Hibiya T (2004) Three-dimensional numerical simulation of M2 internal tides in the East China Sea. J. Geophys. Res. 109: C04027, doi:10.1029/2003JC001923.

[18] Polzin K.L, Toole J.M, Ledwell J.R, Schmitt R.W (1997) Spatial variability of turbulent mixing in the abyssal ocean. Science. 276: 93-96.

[19] Shchepetkin A.F, McWilliams J.C (2005) The Regional Oceanic Modeling System: A split-explicit, free-surface, topography-following-coordinate ocean model. Ocean Modelling. 9: 347-404.

[20] St. Laurent L, Garrett C (2002) The role of internal tides in mixing the deep ocean. J. Phys. Oceanogr. 32: 2882-2899.

[21] St. Laurent L, Simmons H.L, Jayne S.R (2002) Estimating tidally driven mixing in the deep ocean. Geophys. Res. Lett. 29: 2106-2110.

[22] Vlasenko V, Stashchuk N, Hutter K (2005) Baroclinic Tides: Theoretical Modeling and Observational Evidence. Cambridge University Press.

[23] Zilberman N.V, Becker J.M, Merrifield M.A, Carter G.S (2009) Model estimates of M2 internal tide generation over Mid-Atlantic Ridge topography. J. Phys. Oceanogr. 39: 2635-2651.

Chapter 3

Understanding Energy Conservation: Intersection Between Biological and Everyday Life Contexts

Vivien Mweene Chabalengula and Frackson Mumba

Additional information is available at the end of the chapter

http://dx.doi.org/10.5772/51872

1. Introduction

Energy and energy conservation (which is also the First Law of Thermodynamics) are such closely related concepts that it is almost impossible to discuss one without the other in a closed system (Goldring & Osborne, 1994). Energy is commonly defined as the *capacity to do work* (Lee & Liu, 2010), such as driving metabolic reactions and processes (e.g. photosynthesis, muscle contraction) in biological systems. On the other hand, the energy conservation principle is commonly stated as *energy cannot be created or destroyed* (Raven & Johnson, 1999) in almost all science textbooks. However, many science educators argue that this definition may be confusing among students if supporting concepts such as energy transfer, energy flow, and energy transformation (Mclldowie, 1995; Chabalengula et al, 2012) are not incorporated when defining energy conservation. This is because all energy forms are inter-convertible from one form to another without any *loss*; and any apparent loss of energy can be explained as the conversion/transformation of energy into some other form (Raven & Johnson, 1999; Solomon et al, 1993).

The concepts of energy and energy conservation are central scientific ideas which provide an important key to our understanding of the way things happen in the biological, physical, and technological world (Liu, Ebenezer & Fraser, 2002). Furthermore, energy is one of the science concepts that cuts across all science disciplines, and is experienced in our everyday life situations (Saglam-Arslan & Kurnaz, 2009).

However, the synonymous use of the terms *conservation* and *saving* in everyday language usage causes students to misunderstand the scientific meaning of the energy conservation principle. With the current national and global debates on energy saving reality coming to agenda with energy crisis, many students think of energy conservation as energy saving because the latter term is used when issues on depleting energy sources are discussed in media and political realms (Tatar & Oktay, 2007). As such, it is imperative that school-going citizens are scientifically literate about what energy conservation is if they are to delineate energy conservation and energy saving. Therefore we propose that it is vital for science educators to determine students' understanding of energy conservation from a scientific (particularly biological context) and an everyday life context. These two contexts are particularly important because they are interrelated in that biological contexts such as jogging and breathing are always experienced by students on each daily basis so much that they can relate to them very well. Therefore, a compilation of students' understanding of energy conservation based on these two contexts would provide a set of energy conservation data that can be used as a basis for science education curriculum development and instructional design in schools (Lee, 2011).

1.1. Energy conservation principle

According to the energy conservation principle (also known as the first law of thermodynamics), energy cannot be created or destroyed; instead it can be converted from one form to another (Raven & Johnson, 1999; Solomon et al, 1993). All energy forms are inter-convertible from one form to another without any '*loss*'. Any apparent loss of energy can be explained as the conversion of energy into some other form. For example, in biological/living systems, some of the energy is used to drive metabolic processes and some of it dissipates to the atmosphere in the form of heat. Therefore, energy as a conserved quantity at the system level is built upon many supporting concepts such as energy source, energy transfer, energy flow, and energy transformation (Mclldowie, 1995).

The description of energy flow and transformations is frequently used in many biological and technological applications (Ametller & Pinto, 2002; Lin & Hu, 2003). For instance, in biological systems, energy flows from sources such as the sun and moves through a number of carriers to eventual receivers, such as through producers (mostly plants which utilize energy to produce food during photosynthesis) to consumers (animals) in food chains. That is, plant cells transform light energy to chemical energy stored in chemical bonds of food materials. When some herbivores eat the plants, some of this chemical energy is eventually converted to mechanical energy in animals for muscle contraction (useful work), whereas some of it may dissipate to the atmosphere in form of heat energy (Raven & Johnson, 1999). As all these metabolic processes are happening, there is no loss; instead any seemingly loss is explained in terms of energy conversion/transformation. Given the scientific viewpoint of energy conservation especially in biological systems, it is important to remind readers that this concept poses conceptual difficulties among students, as highlighted in the next section.

1.2. Problematic concerns about energy conservation in biology education

As stated earlier on, the energy conservation principle is commonly stated as energy cannot be created or destroyed in most science textbooks. Due to its common appearance in textbooks and its short definition, Tatar and Oktay (2007) point out that energy conservation is one of the most known science principles among students as it is easy to state and remember. As such when asked to state the energy conservation principle, many students tend to recite it with relative easiness, but they are unable to correctly apply it to biological systems. Therefore, the two problematic aspects that make energy conservation a difficult concept to understand among students are:

a. Despite being able to recite it correctly, many students are unable to apply this principle, particularly to biological systems and processes as well as in everyday life situations involving energy (Chabalengula et al, 2012). Part of the reason for this problem has been highlighted by Driver and Warrington (1985, p. 171) who asserted that "very rarely do students consider energy as a conserved quantity; rather it is something that is active for a short period and then disappears".
b. The synonymous use of the terms *conservation* and *saving* in daily life causes students to misunderstand the energy conservation principle (as discussed earlier on).

The two problems stated above led to our motivation to look at the science education literature and determine the extent to which the science education research has tried to remediate the problems. Two issues became evident: First, a review of the previous instruments aimed at diagnosing students' understanding of energy conservation shows that the focus was mainly in physics and engineering contexts rather than in biological contexts. Second, very few studies have used test items which reflected the everyday life situations despite the findings that students have ideas about energy conservation from this perspective. In our opinion, the students' conceptual difficulties with respect to energy conservation are interrelated and connected to other aspects such as everyday life and science discipline contexts, which have not been sufficiently taken into account in previous diagnostic research.

Therefore in this book chapter, we will present findings from the diagnosis we conducted with 90 university biology students' understanding of energy conservation using the pencil and paper test, reflecting test items phrased in the biological and everyday life situations, we developed specifically for this study.

1.3. Previous studies on students' understanding of energy conservation

The science education research has shown that students of all ages have conceptual difficulties and misunderstandings related to the meaning of energy conservation (e.g. Goldring & Osborne, 1994; Pinto et al, 2005), and the application of energy conservation in biological systems (e.g. Barak et al, 1997; Eisten & Stavy, 1988; Fetherston, 1999; Gayford, 1986; Goldring & Osborne, 1994; Kesidou et al, 1993; Kruger et al, 1992; Linjse, 1990; Mann, 2003; Solomon, 1982; Trumper, 1997).

Table 1 shows some previous studies (in chronological order) and the corresponding student errors about energy conservation meanings and applications in biological systems. With respect to what energy conservation means, several researchers have found two common errors: energy conservation means energy saving; and energy conservation is

Some erroneous ideas about energy conservation, as identified by different researchers	Solomon (1982) n=one class (English comprehensive school)	Gayford (1986) n=296 (English A-level biology	Eisten et al (1988) n=188 (Israeli high school & university	Linjse (1990) n=97	Kruger et al (1992) n=159 (English primary school teachers)	Kesidou et al (1993) n=34 (German high school students)	Goldring et al (1994) n=75 (English high school students)	Barak et al (1997) n=104 (76 & 28 Israeli high school	Trumper (1997) n=189 (Israeli pre-service biology	Fetherston (1999) n=94 (Australian high school students)	Mann (2003) n=610 (Australian high school students)	Pinto et al (2005) n=20 (Spanish high school science
Errors relating to meaning of energy conservation												
Energy conservation is synonymous to energy saving							30%					*
Energy conservation is the opposite of energy degradation												*
Errors relating to application of energy conservation in biological systems												
Students' responses to energy-related concepts do not reflect (or if they do, contradict) the idea of energy conservation in biological systems		79%		*	*	100 %	31%	*	*		*	
Energy is **created or formed** during respiration			7%									
Energy is **used up or consumed** during processes in living things				*		29%				15%		
Energy **loss & decrease** during physical activities (e.g. exercises) in living systems are erroneously interpreted as a decrease in its quantity	*											*

Table 1. Summary of students' erroneous ideas about energy conservation
Note: *represents that a corresponding error was observed, but no exact percentages of students holding the error was given.

opposite of energy degradation. With respect to the application of energy conservation, many researchers have found a conceptual failure among students to apply the energy conservation principle (energy cannot be created or destroyed) to biological situations involving energy. That is, the most common errors identified are: energy is created or formed during respiration; energy is used up or consumed during metabolic processes; and energy is lost during physical activities in living organisms. Energy loss and decrease in energy phases are erroneously interpreted as a decrease in its quantity, and not a decrease in energy's usefulness (a correct scientific viewpoint). Scientifically, all energy forms are inter-convertible from one form to another without any loss of energy. Similarly, in all biological processes there is the same amount of energy before as after an event (Starr &Taggart, 1992). However, as shown in Table 1, previous research shows that many students do not realise that energy is conserved in biological processes and systems. Kesidou et al (1993) explain that students often reject the idea of energy conservation because it seems to contradict everyday experiences and language usage where energy is often viewed as being produced and consumed, but not conserved. As such it is important to point out here that some students who seem to have erroneous ideas about energy conservation may be influenced by a language problem, and not a conceptual one. It is possible that whilst students use terms such as *used up* and *lost,* some of them may use these terms figuratively, and not actually conceptualize energy as not being conserved.

1.4. Aims of the chapter

Based on the concerns outlined above, and the previous studies done on students' understanding of energy conservation, the aims of this chapter are three fold: (a) To present the data and results we found on students' understanding of energy conservation in biological and everyday contexts; (b) To develop a diagnostic instrument which can be used to diagnose students' understanding of energy conservation; and (c) To provide suggestions on how biology educators can design their science curriculum and instruction in order to help students understand and apply the concept of energy conservation to biological systems.

The three research questions that guided this study were: (1) Are students able to state the energy conservation principle? (2) To what extent are students able to apply the energy conservation principle to everyday life situations involving biological phenomena? (3) Do students understand what happens to energy during metabolic processes in biological/living organisms?

2. Data collection methods and analysis

2.1. Sample description

The sample consisted of 90 first-year biology students at a South African university. There were 40 males and 50 females. All these students were in a pre-medical program in which they were preparing to go to medical school.

2.2. Development of energy conservation diagnostic test

A survey approach was used for this research, using a pencil-and-paper diagnostic test specifically developed to collect the data. The design of the diagnostic test was based on a process described by Haslam and Treagust (1987). The development of the test involved two main phases: the preparatory and the test formulation phases. The preparatory phase involved three steps. Step 1 involved the drawing up a list of scientifically acceptable propositional knowledge statements about energy conservation and in biological-context. This list was drawn up after interviewing three biology lecturers and then consulting the two prescribed tertiary level textbooks for first-year biology courses at the university where the study was conducted (i.e. Raven & Johnson, 1999; Solomon *et al.*, 1993). The purpose of this step was to draw up the energy conservation conceptual structure that defined this study, and which was used as a guide when constructing the test items, and when marking the answers. The energy conservation conceptual structure we came up with is summarized below:

- According to the first law of thermodynamics (the law of conservation of energy), energy cannot be created or destroyed, but it can be converted from one form to another work.
- All energy forms are inter-convertible from one form to another without any *loss*.
- Any apparent loss of energy can be explained as the conversion of energy into some other form. For example, during each energy conversion in living systems, some of the energy is used to drive the metabolic processes and some of it dissipates to the atmosphere in the form of heat.
- In biological systems, most reactions of organisms involve a complex series of energy transformations. For example, during photosynthesis, plant cells transform light energy to chemical energy stored in chemical bonds of food materials. Some of this chemical energy may eventually be converted to mechanical energy in animals for muscle contraction, if the plant is eaten by an animal.

Step 2 involved compiling a list of common erroneous ideas about energy conservation, reported in the research literature (see Table 1). The purpose of this step was to compile a list of ideas about energy conservation held by students in previous research studies, so these could be included in the diagnostic test to be developed. Step 3 involved defining the content boundaries of the test so that topics which students were expected to know when they enter first year would be included in the test, in order for the test to be used to diagnose prior knowledge as well. The information for this step was acquired through interviews with lecturers who provided data on the prerequisite knowledge required in first year biology.

In the test formulation phase, the diagnostic test had to fulfill two basic criteria: checks basic knowledge about energy conservation; and tests understanding of energy conservation. Testing for understanding is not an easy matter. Various authors have listed criteria that could be used in judging understanding. For example, Sanders and Mokuku (1994) stated that individuals who understand a concept should: know and be able to recognize the name and definition of a concept; be able to define and explain the concept in their own words; be able to recognize instances not previously encountered of the concept; be able to distinguish between and classify instances and non-instances of the concept not previously encountered; and be able to apply the concept to new situations. Through these criteria, it is possible for science educators to determine for sure, their students' understanding level. A number of these criteria were considered when designing the test. For example, the test items were constructed in such a way that students were required to state *energy conservation* in their own words, and to apply their understanding of energy conservation in the closed statements reflecting biological and everyday life scenarios.

2.3. Energy conservation diagnostic test

The test consisted of three questions. Question 1 was an open-ended and aimed at eliciting students' understanding of the energy conservation principle. Actual Question 1 read: *What does the energy conservation principle state?* Question 2 consisted of closed-ended statements phrased in everyday life contexts but reflecting biological phenomena (see Table 3 for the actual statements). These statements were meant to determine the extent to which students are able to apply their understanding of energy conservation in the biological and everyday life situations. The actual Question 2 read: *Below are some statements about energy reflecting everyday life situations and biological phenomena.* Task 1 *(a) For each statement, indicate by placing a tick (✓) in the appropriate box if the statement is scientifically correct or incorrect; (b)Then indicate how sure you are that the answer you have provided is correct by ticking the appropriate box on "Sure", "Think so" or "Guessing".* Task 2 *(a) If the statement is incorrect, underline the word or phrase in the statement which makes it incorrect; and (b) Then write in a word or phrase which would make the statement correct.* Question 3 was open-ended and was meant to elicit students' conceptual understanding of what happens to energy during metabolic processes in living organisms. The actual question 3 read: *Energy is required to drive metabolic processes (such as breathing, muscle contraction) in living organisms. Explain what you think happens to energy during metabolic processes.*

Quality-control steps taken while the instrument was developed included checks on content validity, and rigorous face validation by three university biology experts. The instrument was piloted with 30 first-year biology students who were not involved in the main study. Piloting is an important quality-control procedure as it enables to check on the suitability of individual test items; to gain feedback on how well participants understood the questions, response procedures and instructions so that questions can be improved (Bell, 1987). The pilot group was requested to indicate whether they had problems understanding what they were required to do in each question and each closed statement. They were also further requested to write down the words or phrases which were not clear to them. The results of the pilot showed that majority of the students who indicated having problems suggested their inability to give answers due to lack of knowledge about energy conservation, and not based on misunderstanding the items. As such all test items were maintained.

3. Data analysis

The data were analysed using open-coding and reported as frequency counts and percentages. After a line-by-line analysis of each script, categories and sub-categories to which responses would fit well were developed. The categories which had been developed were given to two biology experts for validation. These experts were asked to check for the following: the scientific correctness of the answers given by students, the appropriateness of the categories developed for the student answers, and whether the students' responses were correctly categorised. This was done by giving each expert ten scripts to go through and code independently. Any differences in the coding were discussed collectively, and a common agreement reached. When individuals rate a product, there is always a possibility that some portion of the agreement between them is due to chance. As such, Cohen (1960) recommends using the kappa statistic to assess interrater agreements involving nominal scale. Cohen's Kappa (k) is a coefficient of interrater agreement that takes into consideration agreement by chance. The interrater agreements between the biology educators on the scientific correctness of the answers provided by students are shown in Table 2.

Test item	Percent agreement	Kappa value
Definitions of energy conservation principle.	96	0.90
Explanations of what happens to energy during metabolic processes.	87	0.85

Table 2. Interrater agreement values

The percentage of agreement ranged from 87 % to 96% with a corresponding kappa coefficients range of 0.85 to 0.90. The percentage agreement of more than 75% and kappa values above 0.5 are considered to indicate good level of interrater agreement (Chiapetta, Fillman & Sethna, 1991). Therefore, the values in this study can be considered good enough to justify reliability. After the categorization process on the scientific correctness of the responses, frequency counts were conducted to get the actual number of students giving each answer, and to calculate the percentages.

4. Results and discussion

The results have been presented along with the discussion. This is because certain aspects of energy such as what happens to energy during metabolic processes need to be discussed alongside the scientifically acceptable perceptions so that the reader(s) can have a clearer perspective as to why the authors of the current study categorized some statements as scientifically correct or incorrect.

4.1. Stating the energy conservation principle

Nearly all students (98%) correctly stated the principle of energy conservation (i.e. energy cannot be created or destroyed). This finding is supported by Tatar et al (2007) who pointed out that this principle is widely known among students due to its easiness in stating and remembering so much that when asked to state it, students do recite it correctly.

However, majority of the students in this study still reverted back to their everyday understanding of energy being *used up*, *created* or *lost* during activity, when they had to apply this principle in closed-ended biological and everyday context statements. This implies that even though these students can correctly state the energy conservation principle, they do not understand it fully so as to apply it to biological situations.

4.2. Application of energy conservation principle in biological situations

Students' understanding and application of energy conservation were elicited using the closed statements that involved biological phenomena presented in everyday life situations, as shown in Tables 3 and 4. Table 3 shows students' responses on the correctness of each statement, and their confidence level of their answers. Table 4 shows students' responses on word/phrase they believed made a statement incorrect, as well as the word/phrase they would write-in to make a statement correct. However, one important point to note as one reads the results in this section is that many students (38%) did not attempt to underline the word or phrase making the statement(s) incorrect, or to write in the word or phrase which would correct the statement(s) which they had indicated were incorrect. The possible validity problem associated with this is that it is not clear whether the students who did not give responses did not follow the instructions, or whether this indicated a lack of understanding of the energy conservation principle. As such, it would have been of value to interview some of these students. However, interviews were not conducted because one of the purposes for this study was to develop a pencil-and-paper diagnostic test that could be used to diagnose students understanding of energy conservation, and which biology educators can easily administer. The specific findings pertaining to students' ability to apply the energy conservation principle are provided in the next subsections.

Test Statements	**How sure are you statement is correct**				**How sure are you statement is incorrect**			
	Statement Correct	Sure	Think so	Guessing	**Statement Incorrect**	Sure	Think so	Guessing
If you go jogging, energy is used up*.	**83 (92%)**	73 (88%)	10 (12%)	0	**6 (7%)**	5 (83%)	1 (17%)	0
After exercise, you can build up your energy levels by resting*.	**39 (43%)**	18 (46%)	17 (44%)	4 (10%)	**50 (56%)**	17 (34%)	32 (64%)	0

When you are asleep your body does not require any energy because it is not active*.	**6 (7%)**	4 (67%)	2 (33%)	0	**83 (92%)**	71 (86%)	9 (11%)	0
During exercise, energy is built up in the body*.	**14 (16%)**	6 (43%)	7 (50%)	0	**74 (82%)**	47 (64%)	23 (31%)	3 (4%)
When living things are active, they lose energy*.	**53 (59%)**	30 (57%)	18 (34%)	3 (6%)	**36 (40%)**	20 (56%)	15 (42%)	0

Table 3. Students' responses on correctness of statements, and how sure they were about their answers.
Notes: Figures outside the brackets represent the actual number of students who responded.
* Statement is scientifically incorrect.

Number of students who did not attempt the test item 34 (38%)								
Number of students who responded to the test item 56 (62%)								
	Task 1				**Task 2**			
Statements reflecting biological phenomena	**Underlined word or phrase which makes statement incorrect**	Actual number of students	% of who responded	% of whole	**Written-in Word or phrase which makes statement correct**	Number of students	% of who responded	% of whole sample
If you go jogging, energy is used up*	Incorrect phrase -used up	4	7	4	Acceptable - converted to different forms	3	5	3
					Unacceptable -lost	1	2	1
After exercise, you can build up your energy levels by resting*	Incorrect phrase -build up	9	16	10	Acceptable -do not build up	4	7	4
	Incorrect word -resting	22	39	25	Acceptable -eating(food consumption)	15	27	16
					Unacceptable -stop using energy	3	6	4

When you are asleep your body does not require any energy because it is not active*	Incorrect phrase -does not require energy because it is not active	47	84	52	Acceptable -require energy as it is still active - many body processes (e.g. respiration) still occur	21 8	37 14	23 9
During exercise, energy is built up in the body*	Incorrect phrase -built up	42	75	47	Acceptable - converted to different forms Unacceptable - used up - lost	7 23 7	12 41 13	8 26 8
When living things are active, they lose energy*	Incorrect word -lose	25	45	28	Acceptable -convert it to different forms Unacceptable -use up	7 10	12 18	7 11

Table 4. Students' responses on words/phrases making the statements incorrect, and how they would correct them.
Note: * Statement is scientifically incorrect

4.2.1. Ideas relating to energy being used up when an organism is active

The idea that energy is *used up* during activities was erroneously accepted by majority of the students. That is, 92% of them erroneously indicated that the statement *If you go jogging, energy is used up,* is correct and most of these students (73 of them) were sure their response was correct. To the contrary, only 6 students correctly indicated that the statement is incorrect. When asked to underline the phrase making the statement incorrect, only four of the six students (who indicated that the statement was incorrect) correctly underlined the phrase *used up*. Three of these students correctly wrote in the phrase "energy is converted into different forms". One student wrote in the word *lost,* however without the added explanation that heat energy is lost to the body during respiration, this answer has to be judged as erroneous.

The idea that energy is *used up* during activities and in processes is documented by many researchers (e.g. Linjse, 1990; Kesidou & Duit, 1993; Fetherston, 1999). For example, Kesidou et al found that 29% of the students gave an explanation in which they explicitly employed the ideas of energy being used up. In another study by Fetherston (1999), a much smaller percentage of students (15%) erroneously stated that energy is *used up* during processes in living things. A similar finding was documented by Kesidou et al (1993) and Fetherston (1999) in which high school students had an erroneous view that energy is used up.

Scientifically, energy cannot be used up, instead, it remains constant despite the energy changes which occur, according to the First Law of Thermodynamics. However, in everyday language usage, energy is viewed as a substance which becomes used up in situations dealing with activity such as exercises. Many students may erroneously conclude that "energy is used up, as in batteries which go flat" or "when food is eaten" or during activities (Kesidou et al, 1993; Fetherston, 1999).

4.2.2. *Ideas relating to energy being built up when an organism is at rest or active*

Two statements tested the idea that energy can be *built up*. Almost half of the students in the sample (43%) erroneously indicated that the statement *After exercise, you can build up your energy levels by resting* is correct, although 17 of them had doubts about whether their answer was correct. However, quite a large number of the students (56% of them) correctly indicated that the statement was scientifically incorrect; although very few of them (17) were sure their response was correct. The majority (32) seemed to be unsure that their answer was correct. When asked to underline the word/phrase that made the statement incorrect, nine students correctly underlined the phrase "build up" as being erroneous. However, when it came to writing-in the word/phrase to correct the statement, very few students did so. For instance, only one student correctly wrote-in the phrase "do not build up" to correct the statement.

The statement *During exercise, energy is built up in the body,* also tested whether students agree that energy is "built up". And 16% of them erroneously indicated that it was correct, and six of them were sure of their answer. To the contrary, almost all students (82%) held a scientists' view that the statement was incorrect, and more than half of these students (47 of them) were sure of their answer. When asked to underline the incorrect word/phrase in the statement, 42 students recognised and underlined "build up" as an erroneous phrase. However, only 7 students wrote-in an acceptable phrase "converted to different forms" to correct the statement. The rest of the students still provided unacceptable phrases such as *used up* (23 students) and *lost* (7 students). As explained earlier, the word *lost* contradicts the principle of energy conservation, if there is no added explanation that heat energy is lost to the body during respiration.

4.2.3. *Ideas relating to energy not being required when a body is at rest*

This claim was tested in the statement *When you are asleep, your body does not require any energy because it is not active.* Very few students (7%) erroneously stated that the statement was correct. To the contrary, nearly all students (92%) correctly identified the statement as scientifically incorrect, and 71 of them were sure their response was correct. When asked to underline the word/phrase which makes the statement wrong, 47 students correctly underlined "does not require any energy because it is not active". Of the 47 students, 21 of them wrote in an acceptable phrase "does require energy as it is still active" and 8 other students wrote in another acceptable phrase "many body processes (e.g. respiration) still occur" to correct the statement. Most students in this study appeared to hold a scientifically acceptable view that the human body is active even at rest, and therefore energy is always required.

4.2.4. *Ideas relating to energy being lost when an organisms is active*

Slightly more than half of the students in the sample (59%) erroneously indicated that the statement *When living things are active, they lose energy* was correct, and 30 of them were sure of their answer. On the other hand, 40% of them exhibited a scientific view that the statement was incorrect, and more than half (20 of them) were sure of their answer. When asked to underline an incorrect word in the statement, 25 students correctly underlined the word "lose", with 7 students writing in the acceptable phrase "convert or transform energy into different forms". However, 10 of 25 students who underlined the word "lose" erroneously replaced it with an unacceptable phrase "used up".

The idea that *energy is lost* during activity in living organisms was documented by Solomon (1982). In everyday life experiences, people may well consider energy to be lost, and that the amount of energy decreases, especially during and after exercises. As a result, the scientific viewpoint that the amount of energy remains constant, despite the transformations that occur (i.e. energy conservation) does not seem to be applied in everyday experiences. According to the First Law of Thermodynamics (the law of conservation of energy), energy cannot be created or destroyed, but it can be converted from one form to another without any *loss;* any apparent loss of energy can be explained as the conversion of energy into some other form (Solomon *et al.,* 1993; Raven and Johnson, 1999). For example, during each energy conversion in living systems, some of the energy is used to drive the metabolic processes and some of it dissipates to the atmosphere in the form of heat. When talking about energy, the idea of energy being *lost* usually appears to be erroneous because it implies that energy is not conserved, although energy in the form of heat certainly *leaves* the body.

4.2.5. *Inability to apply the energy conservation principle to biological situations*

Although nearly all students (98%) correctly stated the energy conservation principle (energy cannot be created or destroyed) in test question 1, their ability to apply this concept to biological situations proved otherwise, as shown in Tables 3 & 4. The contradictory answers given to the closed statements could suggest that the students are unable to apply the idea of energy conservation. This situation may suggest that students could have rote-learned this principle or could have seen it in textbooks. Another source for this problem could be the language differences between science and everyday usage. Using phrases such as *used up, build up* or *lost* are not scientifically correct when talking about energy conservation.

The inability to apply the scientific idea of energy conservation in biological systems among students has also been identified by other researchers (e.g. Barak et al., 1997; Gayford, 1986; Goldring & Osborne, 1994; Kesidou & Duit, 1993; Kruger et al., 1992; Linjse, 1990; Liu et al,

2002; Solomon, 1985; Trumper, 1997; Warren, 1986). For instance, Barak et al (1997) conducted a study in which they found that majority of the students had difficulties in applying the law of energy conservation in a biological context. Other researchers such as Kesidou et al (1993) found that none of their 34 Grade Ten German students could apply the idea of energy conservation. Similarly, even teachers have been found to have this problem as documented by Kruger et al (1992) where majority of the answers teachers provided contradicted with the principle of energy conservation.

In light of students' inability to apply the concept of energy conservation, Warren (1986) explained that the *conservation of energy* is problematic because it implies *saving fuel* when used in everyday life and social understanding. In another explanation, Duit (1981) argued that energy conservation may not be intelligible to most learners because in everyday usage, energy is viewed to be "produced and consumed, but not conserved" (p. 292). The inability of students in this study to consistently apply this principle may suggest that students could have rote-learned this law or could have seen it in textbooks.

4.3. What happens to energy during metabolic processes in living organisms

Students' responses on what happens to energy during metabolism are in Table 5.

What happens to energy during metabolic processes	**% of Students**
Statements to do with energy being created or degraded	**55**
Scientifically acceptable statement	40
Energy is neither created nor destroyed	9
Scientifically unacceptable statements	6
Energy is created/made during processes (e.g. photosynthesis & respiration) in organisms	
Energy is degraded during processes	
Statements to do with energy used up & lost during processes	**37**
Scientifically unacceptable statements	
	27
Energy is used up during processes (e.g. respiration) in living organisms	10
Energy is lost during an activity (e.g. exercises) in organisms	
Statements to do with energy being transformed or transferred	**7**
Scientifically acceptable statement	7
Energy can be transformed or converted from one form to another	

Table 5. Students' responses on what happens to energy during metabolic processes

Lee, H. & Liu, O. L. (2010). Assessing learning progression of energy concepts across Middle school grades: the knowledge integration perspective. *Science Education,* Vol,94, No, 4, (July 2010), pp. 665- 688. ISSN: 1098-237X.

Lin, C. Y. & Hu, R. (2003). Students 'understanding of energy flow and matter cycling in the context of the food chain, photosynthesis, and respiration. *International Journal of Science Education,* Vol. 25, No. 12, pp. 1529-1544, ISSN 0950-0693.

Linjse, P. (1990). Energy between the life-world of pupils and the world of physics. Journal of Science Education, Vol. 74, No. 1, pp. 571-583. ISSN 0950-0693.

Liu, X., Ebenezer, J. & Fraser, D. M. (2002). Structural characteristics of university engineering students' conceptions of energy. *Journal of Research in Science Teaching, 39*(5), 423–441. Vol. 39, No. 5, pp. 423-441, ISSN: 1098-2736.

Mann, M. F. (2003). Students' use of formal and informal knowledge about energy and the human body. Unpublished PhD Thesis. Curtin University of Technology, Australia.

McIldowie, E. (1995). Energy transfer. Where did we go wrong? *Physics Education,* Vol. 30, No. 4, (July 1995), pp. 228- 230, ISSN 0031-9120.

Pinto, R., Couso, D. & Gutierrez, R. (2005). Using Research on Teachers' Transformations of Innovations to Inform Teacher Education. The Case of Energy Degradation. *Science Education,* Vol, 89, No. 1, (January 2005), pp. 38-55, ISSN: 1098-237X.

Raven, P.H. & Johnson, G.B. (1999). *Biology* (Fifth edition), McGraw-Hill, ISBN-10: 0697353532 , Boston.

Saglam-Arslan, A. & Kurnaz, M, A. (2009). Prospective physics teachers' level of understanding energy, power and force concepts. *Asia-Pacific Forum on Science Learning and Teaching,* Vol. 10, No. 1, (June 2009), Article 6, ISSN 1609-4913.

Sanders, M. & Mokuku, T. (1994). How valid is face validity? *Proceedings of the Southern African Association for Research and Development in Mathematics and Science Education,* pp. 479-489, University of Durban-Westville, Durban, South Africa, January 1994.

Solomon, J. (1982). How children learn about energy or does the first law come first? *School Science Review,* Vol. 63, No. 224, pp. 415-422.

Solomon, J. (1985). Teaching the conservation of energy. *Physics Education, Vol.* 20, No. 4, (July 1985), pp. 165-170, ISSN 0031-9120.

Solomon, E.P; Berg, L.R; Martin, D.W. and Villee, C. (1993). *Biology* (Third edition), Saunders College Publishers, ISBN 0030974992, New York.

Starr, C. &Taggart, R. (1992). *Biology: The unity and diversity of life* (6th edition), Wadsworth Publishing Company, *ISBN* 0-7167-3126-6, Belmont, California.

Tatar, E. & Oktay, M. (2007). Students' Misunderstandings about the Energy Conservation Principle: A General View to Studies in Literature. *International Journal of Environmental and Science Education,* Vol. 2, No. 3, pp. 79-81, ISSN 1306-3065.

Trumper, R. (1997). A survey of conceptions of energy of Israeli pre-service high school biology teachers. *International Journal of Science Education,* Vol. 19, No. 1, pp. 31-46, ISSN 0950-0693.

Warren, J. (1986). At what stage should energy be taught? *Physics Education,* Vol. 21, No. 3, (May 1986), pp. 154-156, ISSN 0031-9120.

Section 2

Applications of Energy Conservation

Chapter 4

Hydro Power

Mohammed Taih Gatte and Rasim Azeez Kadhim

Additional information is available at the end of the chapter

http://dx.doi.org/10.5772/52269

1. Introduction

Humans have used the power of flowing water for thousands of years. Early civilizations used wooden paddle wheels to grind corn and wheat to flour. The word *Hydro* comes from the Greek word for water. *Hydropower* traditionally represents the energy generated by damming a river and using turbine systems to generate electrical power. However, there are several other ways we can generate energy using the power of water. Ocean waves, tidal currents and ocean water temperature differences can all be harnessed to generate energy. More than 70 percent of the earth is covered by water. The United States is one of the worlds top producers of hydropower (see chart). As much as 12 percent of the electrical energy generated in the U.S. is currently derived from hydropower systems. Parts of the Pacific Northwest generate as much as 70 percent of their electricity using hydroelectric sources. More than half the renewable energy generated in the United States comes from hydroelectric dams. Hydroelectric power is currently the least expensive source of electrical power and is much cleaner than power generated using fossil fuels.

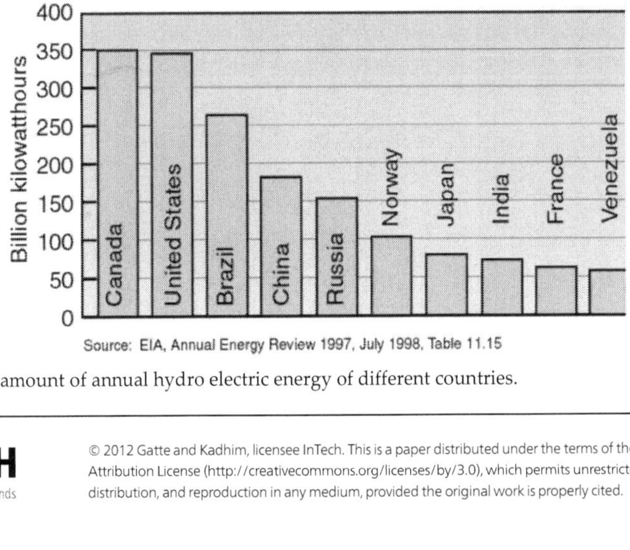

Figure 1. The amount of annual hydro electric energy of different countries.

Flowing and falling water have potential energy. Hydropower comes from converting energy in flowing water by means of a water wheel or through a turbine into useful mechanical power. This power is converted into electricity using an electric generator or is used directly to run milling machines. The potential energy of water may be used directly without conversion operation because of different in elevation diverted the water through a pipelines in order to supply the water in the daily usage .

2. History of hydro power

In the ancient times waterwheels were used extensively, but it was only at the beginning of the 19th Century with the invention of the hydro turbines that the use of hydropower got popularized. Small-scale hydropower was the most common way of electricity generating in the early 20th century. The first commercial use of hydroelectric power to produce electricity was a waterwheel on the Fox River in Wisconsin in 1882 that supplied power for lighting of two paper mills and a house. Within a matter of weeks for this installation, a power plant was also put into commercial service at Minneapolis1. India has a century old history of hydropower and the beginning was from small hydro. The first hydro power plant was of 130 kW set up in Darjeeling during 1897, marked the development of hydropower in the country. Similarly, by 1924 Switzerland had nearly 7000 small scale hydropower stations in use. Even today, Small hydro is the largest contributor of electricity from renewable energy sources, both at European and world level. With the advancement of technology, and increasing requirement of electricity, the thrust of electricity generation was shifted to large size hydro and thermal power stations.

However, it is only during the last two decades that there is a renewed interest in the development of small hydro power (SHP) projects mainly due to its benefits particularly concerning environment and ability to produce power in remote areas. Small hydro projects are economically viable and have relatively short gestation period. The major constraints associated with large hydro projects are usually not encountered in small hydro projects. Renewed interest in the technology of small scale hydropower actually started in China which has more than 85,000 small-scale, electricity producing, hydropower plants.

Hydropower will continue to play important role throughout the 21st Century, in world electricity supply. Hydropower development does have some challenges besides the technical, economic and environmental advantages it shares above other power generation technologies.

At the beginning of the new Millennium hydropower provided almost 20% (2600 TWh/year) of the electricity world consumption (12900 TWh/year). It plays a major role in several countries. According to a study of hydropower resources in 175 countries, more than 150 have hydropower resources. For 65 of them, hydro produces more than 50% of electricity; for 24, more than 90% and 10 countries have almost all their electricity requirements met through hydropower.

3. Hydro power system classification

A different countries have different criteria to classify hydro power plants, a general classification of hydro power plants is as follows in table 1 :

Type	*Capacity in KW*
Micro Hydro	*Up to 100*
Mini Hydro	*101 to 2000*
Small Hydro	*2001 to 25000*
Large Hydro	> 25000

Table 1. The hydro power types according to output power

The hydro plants are also classified according to the "Head" or the vertical distance through which the water is made to impact the turbines. The usual classifications are given in table 2. below:

Type	*Head Range*
High head	*100 m and above*
Medium head	*30 – 100 m*
Low head	2 – 30 m

Table 2. The hydro power types according to head

These ranges are not rigid but are merely means of categorizing sites. Schemes can also be defined as:

- Run-of-river schemes
- Schemes with the powerhouse located at the base of a dam
- Schemes integrated on a canal or in a water supply pipe

Most of small hydro power plants are "run-of-river" schemes, In order to imply that they do not have any water storage capability. The power is generated only when enough water is available from the river/stream. When the stream/river flow reduces below the design flow value, the generation ceases as the water does not flow through the intake structure into the turbines. Small hydro plants may be stand alone systems in isolated areas/sites, but could also be grid connected (either local grids or regional/national grids). The connection to the grid has the advantage of easier control of the electrical system frequency of the electricity, but has the disadvantage of being tripped off the system due to problems outside of the plant operator's control.

4. The hydro power principles

Power generation from water depends upon a combination of head and flow. Both must be available to produce electricity. Water is diverted from a stream into a pipeline, where it is directed downhill and then through the turbine (flow). The vertical drop (head) creates pressure at the bottom end of the pipeline. The pressurized water emerging from the end of the pipe creates the force that drives the turbine. The turbine in turn drives the generator where electrical power is produced. More flow or more head produces more electricity. Electrical power output will always be slightly less than water power input due to turbine and system inefficiencies. Water pressure or Head is created by the difference in elevation between the water intake and the turbine. Head can be expressed as vertical distance (feet or meters), or as pressure, such as pounds per square inch (psi). Net head is the pressure available at the turbine when water is flowing, which will always be less than the pressure when the water flow is turned off (static head), due to the friction between the water and the pipe. Pipeline diameter also has an effect on net head. Flow is quantity of water available, and is expressed as 'volume per unit of time', such as gallons per minute (gpm), cubic meters per second (m^3/s), or liters per minute (lpm). Design flow is the maximum flow for which the hydro system is designed. It will likely be less than the maximum flow of the stream (especially during the rainy season), more than the minimum flow, and a compromise between potential electrical output and system cost. The theoretical power (P) available from a given head of water is in exact proportion to the head and the quantity of water available.

$$P = Q \times H \times e \times 9.81 \left(kW\right)$$

Where

P Power at the generator terminal, in kilowatts (kW)

H The gross head from the pipeline intake to the tail water in meters (m)

Q Flow in pipeline, in cubic meters per second (m^3/s)

e The efficiency of the plant, considering head loss in the pipeline and the efficiency of the turbine and generator, expressed by a decimal (e.g. 85% efficiency= 0.85)

9.81 is a constant and is the product of the density of water and the acceleration due to gravity (g)

Example: A site has a head of 100 m with flow of 0.1 m^3/s .calculate the output power output if a) e=100% , b) e=50% .

Solution:

a. P = 100 * 0.1 * 9.81 * 100% = 98.1 KW
b. P = 100 * 0.1 * 9.81 * 50% = 49 KW

Table 3. below shows the output power (KW) with (e = 50%) for different heads and flows

Head (m)	Flow Rates (m^3/s)								
	0.01	0.02	0.04	0.08	0.1	0.2	0.5	1.0	2.0
	Output Power in (KW)								
1	4.9	9.8	19.6	39.2	49	98	245	490	980
2	9.8	19.6	39.2	78.4	98	196	490	980	1960
4	19.6	39.2	78.4	156.8	196	392	980	1960	3920
8	39.2	78.4	156.8	313.6	392	784	1960	3920	7840
10	49	98	196	392	490	980	2450	4900	9800
15	73.5	147	294	588	735	1470	3675	7350	14700
20	98	196	392	784	980	1960	4900	9800	19600
30	147	294	588	1176	1470	2940	7350	14700	29400
40	196	392	784	1568	1960	3920	9800	19600	39200
50	245	490	980	1960	2450	4900	12250	24500	49000
100	490	980	1960	3920	4900	9800	24500	49000	98000
150	735	1470	2940	5880	7350	14700	36750	73500	147000
200	980	1960	3920	7840	9800	19600	49000	98000	196000

Table 3. Table 3. the output hydro power with different heads and flow rates

5. Basic components of a hydropower system

Figure 2 below shows the major components of a typical hydropower scheme. The water in the river is diverted by the weir through an opening in the river side (the 'intake') into a channel (this could be open or buried depending upon the site conditions). A settling basin is built in to the channel to remove sand and silt from the water. The channel follows the contour of the area so as to preserve the elevation of the diverted water. The channel directs the water into a small reservoir/tank known as the 'forebay' from where it is directed on to the turbines through a closed pipe known as the 'penstock'. The penstock essentially directs the water in a uniform stream on to the turbine at a lower level. The turning shaft of the turbine can be used to rotate a mechanical device (such as a grinding mill, oil expeller, wood lathe, etc.) directly, or to operate an electricity generator. The machinery or appliances which are energized by the turbine are called the 'load'. When electricity is generated, the 'power house' where the generator is located transfers the electricity to a step-up 'transformer' which is then transmitted to the grid sub-station or to the village/area where this electricity is to be used.

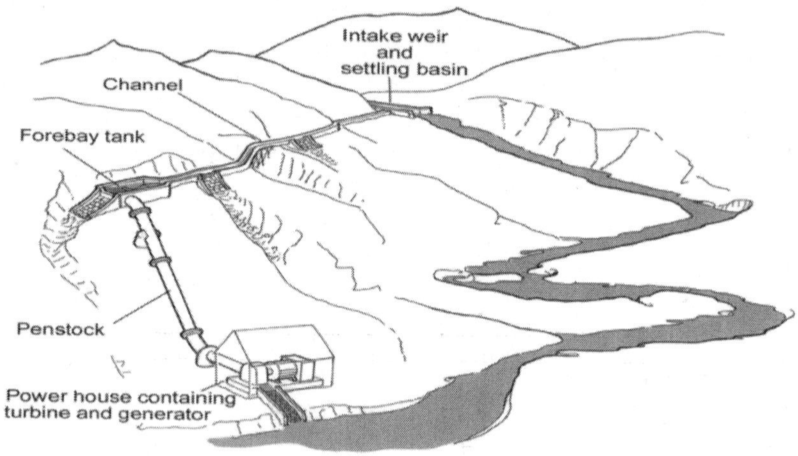

Figure 2. The hydro power plant components

5.1. Civil works

Civil works structures control the water that runs through a hydropower system, and conveyances are a large part of the project work. It is important that civil structures are located in suitable sites and designed for optimum performance and stability. Other factors should be considered in order to reduce cost and ensure a reliable system, including the use of appropriate technology, the best use of local materials and local labour, selection of cost-effective and environmentally friendly structures, landslide-area treatment and drainage-area treatment. Head works consist of the weir (see Figure 2), the water intake and protection works at the intake to safely divert water to the headrace canal. At some sites you may be able to install the penstock directly in the intake, with no need for a canal.

A hydropower station essentially needs water to be diverted from the stream and brought to the turbines without losing the elevation/head. Given below are some of the important factors that must be kept in mind while designing a hydropower system: **Available head:** The design of the system has effects on the net head delivered to the turbine. **Flow variations:** The river flow varies during the year but the hydro installation is designed for almost a constant flow. **Sediment:** Flowing water in the river sometimes carry small particles of hard abrasive matter (sediment) which can cause wear to the turbine if they are not removed before the water enters the penstock. **Floods**: Flood water will carry larger suspended particles and will even cause large stones to roll along the stream bed. **Turbulence:** In all parts of the water supply line, including the weir, the intake and the channel, sudden alterations to the flow direction will create turbulence which erodes structures and causes energy losses. Most common civil structures used in a hydro power scheme are:

5.1.1. Weir and intake

A hydro power system necessitates that water from the river to be diverted and extracted in a reliable and controllable manner. The water flowing in the channel must be regulated during high river flow and low flow conditions. A weir can be used to raise the water level and ensure a constant supply to the intake. Sometimes it is possible to avoid building a weir by using natural features of the river. A permanent pool in river could also act as a weir.

Another condition in site selection of the weir is to protect it from damage. Sometimes, in remote hilly regions, where annual flooding is common it may be prudent to build temporary weir using local resources and manpower. The temporary weir is a simple structure at low cost using local labour, skills and materials. It is expected to be destroyed by annual or bi-annual flooding. However, advanced planning has to be done for rebuilding of the weir.

The intake of a hydro power is designed to divert only a portion of the stream flow or the complete flow depending upon the flow conditions and the requirement. Hydro power schemes use different types of intakes distinguished by the method used to divert the water into the intake. For hydro power schemes, intake systems are smaller and simpler. The following three types of intakes have been described here: side intake with and without a weir and the bottom intake. The advantages and disadvantages associated with each of these are given in the table below:

Attributes	**Side intake without weir**	**Side intake with weir**	**Bottom intake**
Advantages	1. Relatively cheap 2. No complex machinery required for construction	1. Control over water level 2. Little maintenance necessary (if well designed)	1. Very useful at fluctuating flows. Even the lowest flow can be diverted. 2. No maintenance required (if well designed)
Dis-advantages	1. Regular maintenance and repairs required. 2. At low flows very little water will be diverted and therefore this type of intake is not suitable for rivers with great fluctuations in flow.	1. Low flow cannot be diverted properly 2. Modern materials like concrete necessary.	1. Expensive. 2. Local materials not suitable. 3. Good design required to prevent blockage by sediment.

Table 4. The advantages and disadvantages of the intake types

5.1.2. *Power channels*

The power channel or simply a channel conducts the water from the intake to the forebay tank. The length of a channel depends upon the topography of the region and the distance of powerhouse from the intake. Also the designing of the MHP systems states the length of the channel sometimes a long channel combined with a short penstock can be cheaper or required, while in other cases a combination of short channel with long penstock would be more suitable.

In the Himalayan region, the hydro power channels are sometimes as long as a few kilometers to create a head of 10 to 60 meters or more. Generally power channels are excavated and to reduce friction and prevent leakages these are often lined with cement, clay or polythene sheet. Size and shape of a channel and material used for lining are often a dictated by cost and head considerations. During the process of flowing past the walls and bed material, the water loses energy. The rougher material have greater friction loss and higher elevation difference needed between channel entry and exit.

In hilly regions it is common that the power channel would have to cross small streams. In such situations it is often prudent to build a complete crossing over the channel, during rainy season and flash floods, rocks/mud may block the channel or wash away sections of the channel. Sometimes just the provision of a drain running under the channel (in case of very small streams along stable slopes) is usually adequate. The power channel has some important parts which are described in the sub-sections below:

5.1.3. *Settling basin*

The water diverted from the stream and carried by the channel usually carries a suspension of small particles such as sand that are hard and abrasive and can cause expensive damage and rapid wear to turbine runners. To get rid of such particles and sediments, the water flow is allowed to slow down in 'settling basins' so that the sand and silt particles settle on the basin floor. The deposits are then periodically flushed. The design of settling basin depends upon the flow quantity, speed of flow and the tolerance level of the turbine (smallest particle that can be allowed). The maximum speed of the water in the settling basin can thus be calculated as slower the flow, lower is the carrying capacity of the water. The flow speed in the settling basin can be lowered by increasing the cross section area.

5.1.4. *Spillways*

Spillways along the power channel are designed to permit overflow at certain points along the channel. The spillway acts as a flow regulator for the channel. During floods the water flow through the intake can be twice the normal channel flow, so the spillway must be large enough to divert this excess flow. The spillway can also be designed with control gates to empty the channel. The spillway should be designed in such a manner that the excess flow is fed back without damaging the foundations of the channel.

Figure 3. Spillway of Darbandikhan dam

5.1.5. Forebay tank

The forebay tank serves the purpose of providing steady and continuous flow into the turbine through the penstocks. Forebay also acts as the last settling basin and allows the last particles to settle down before the water enters the penstock. Forebay can also be a reservoir to store water depending on its size (large dams or reservoirs in large hydropower schemes are technically forebay).

A sluice will make it possible to close the entrance to the penstock. In front of the penstock a trashrack need to be installed to prevent large particles to enter the penstock. A spillway completes the forebay tank.

Figure 4. Forebay tank

5.2. Penstock

The penstock pipe transports water under pressure from the forebay tank to the turbine, where the potential energy of water is converted into kinetic energy in order to rotate the turbine. The penstock is often the most expensive item in the project budget – as much as 40 percent is not uncommon in high-head installations. It is therefore worthwhile to optimize its design in order to minimize its cost. The choice of size and type of penstock depends on several factors that are explained briefly in this section. Basically, the trade-off is between head loss and capital cost.

Head loss due to friction in the penstock pipe depends principally on the velocity of water, the roughness of pipe wall and the length and diameter of pipe. The losses decrease substantially with increased pipe diameter. Conversely, pipe costs increase steeply with the increase of diameter. Therefore, a compromise between cost and the required performance. The design philosophy is first identify available pipe options, select a target head loss of 5 to 10 percent or less of the gross head, and keep the length as short as possible. Several options for sizes and types of materials may need to be calculated and evaluated in order to find a suitable penstock pipe. A smaller penstock may save on capital costs, but the extra head loss may account for lost energy and revenue from generated electricity (if you are selling the power). In smaller systems, the allowable head loss can be as much as 33 percent. This is particularly relevant to developers who combine domestic water supply and penstock in the same pipe.

Several factors should be considered when deciding which material will be used in a particular penstock: design pressure, the roughness of the pipe's interior surface, method of joining, weight and ease of installation, accessibility to the site, design life and maintenance, weather conditions, availability, relative cost and likelihood of structural damage.

The pressure rating of the penstock is critical because the pipe wall must be thick enough to withstand the maximum water pressure; otherwise there will be a risk of bursting. The pressure of the water in the penstock depends on the head; the higher head is higher pressure. Pressure ratings are normally given in bar units or PSI; 10.2 m of head will exert a pressure of 1 bar, or 14.5 PSI. The penstock becomes more expensive as the pressure rating increases.

The most commonly used materials for a penstock are HDPE, uPVC and mild steel because of their suitability, availability and affordability. Layout of the penstock pipelines depends on their material, the nature of terrain and environmental considerations; they are generally surface-mounted or buried underground. Special attention is necessary where a penstock is installed in a very cold environment; protection from ice and frost must be considered. In severe frost areas, penstocks should always be buried below the frost line. Where freezing is not a concern, the penstock may be left above ground. However, it is generally preferable to bury the penstock to provide protection from expansion, animals and falling trees. Because of changes in the ambient temperature, the length of the penstock pipe may be subjected to expansion and contraction. Expansion joints are used to compensate for maximum possible changes in length.

Material	*Friction*	*Weight*	*Corrosion*	*Cost*	*Jointing*	*Pressure*
Mild Steel	XXX	XXX	XXX	XXXX	XXXX	XXXXX
HDPE[2]	XXXXX	XXXXX	XXXXX	XX	XX	XXXXX
uPVC[3]	XXXXX	XXXXX	XXXX	XXXX	XXXX	XXXXX

[1]Adapted from Frankel, Peter , et al. Micro-Hydro Power : A Guide for development Workers . London, U.K: Intermediate Technology Publications in association with the Stockholm Environment Institute , 1991.
X = Poor ------------ XXXXX = Excellent
[2]HDPE = High density polyethylene
[3]uPVC = Unplastified polyvinyl chroride

Table 5. Comparison of Penstock Materials[1]

The hydrostatic pressure created from the head must be determined so that a suitable wall thickness can be determined. This pressure is given by Equation (1).

$$Pressure = p*g*H \tag{1}$$

p density of water
g acceleration due to gravity
H head

With the pressure calculated, the minimal wall thickness can then be calculated from Equation (2).

$$t = P*D / 2*s \tag{2}$$

t wall thickness
P hydrostatic pressure
D diameter
s allowable tensile stress

Friction is always present, even in fluids, it is the force that resists the movement of objects. When you move a solid on a hard surface, there is friction between the object and the surface. If you put wheels on it, there will be less friction. In the case of moving fluids such as water, there is even less friction but it can become significant for long pipes. Friction can be also be high for short pipes which have a high flow rate and small diameter as in the syringe example. In fluids, friction occurs between fluid layers that are traveling at different velocities within the pipe (see Figure 5). There is a natural tendency for the fluid velocity to be higher in the center of the pipe than near the wall of the pipe. Friction will also be high for viscous fluids and fluids with suspended particles.

Another cause of friction is the interaction of the fluid with the pipe wall, the rougher pipe has higher friction. Friction depends on:

- average velocity of the fluid within the pipe
- viscosity
- pipe surface roughness

the increase in any one of these parameters will increase friction. The amount of energy required to overcome the total friction loss within the system has to be supplied by the pump if you want to achieve the required flow rate. In industrial systems, friction is not normally a large part of a pump's energy output. For typical systems, it is around 25% of the total. If it becomes much higher then you should examine the system to see if the pipes are too small. However all pump systems are different, in some systems the friction energy may represent 100% of the pump's energy, this is what makes pump systems interesting, there is a million and one applications for them. In household systems, friction can be a greater proportion of the pump energy output, maybe up to 50% of the total, this is because small pipes produce higher friction than larger pipes for the same average fluid velocity in the pipe . Another cause of friction are the fittings (elbows, tees, y's, etc) required to get the fluid from point to point. Each one has a particular effect on the fluid streamlines. For example in the case of the elbow , the fluid streamlines that are closest to the tight inner radius of the elbow lift off from the pipe surface forming small vortexes that consume energy. This energy loss is small for one elbow but if you have several elbows and other fittings it can become significant. Generally, they rarely represent more than 30% of the total friction due to the overall pipe length.

To sum up, head losses in a penstock depend on:

- Its shape: singularities as elbows or forks tend to increase head losses
- Its internal diameter
- Its wall roughness and its evolution due to its degradation or/ and to wall deposits.

It may be recalled here that energy loss due to friction in a penstock can be estimated as being inversely proportional to its diameter to the power of five. For instance, a diameter increase of 20% leads to a head losses decrease of 60%. Figure 5. shows the relation between the head loss (feet) with the flow rate (gpm) for a 100 feet PVC class 160 plastic pipe.

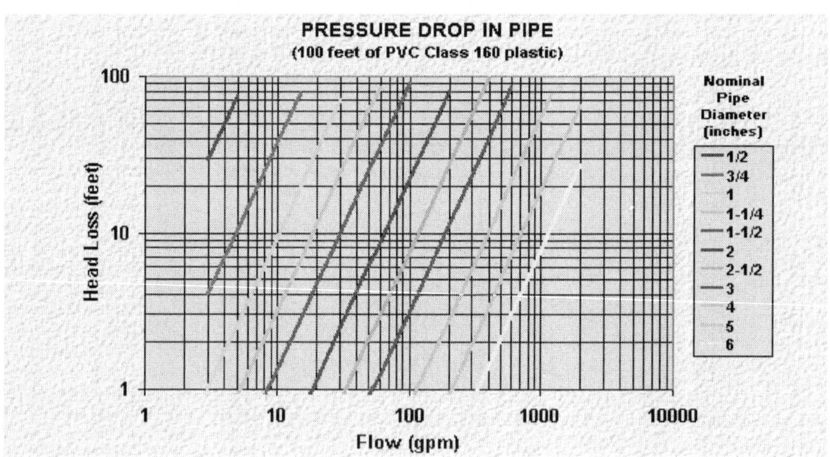

Figure 5. The relation of head loss with the flow

5.3. Turbines

Turbine is the main piece of equipment in the hydro power scheme that converts energy of the falling water into the rotating shaft power. The selection of the most suitable turbine for any particular hydro site depends mainly on two of the site characteristics – head and flow available. All turbines have a power-speed characteristic. This means they will operate most efficiently at a particular speed, head and flow combination. Thus the desired running speed of the generator or the devices being connected/ loading on to the turbine also influence selection. Other important consideration is whether the turbine is expected to generate power at part-flow conditions. The design speed of a turbine is largely determined by the head under which it operates. Turbines can be classified as high head, medium head or low head machines. They are also typified by the operating principle and can be either impulse or reaction turbines. The basic turbine classification is given in the table below:

Turbine Runner	*High Head*	*Medium Head*	*Low Head*	*Ultra- Low Head*
Impulse	**Pelton** **Turgo**	**Cross- flow** **Turgo** **Muti-jet Pelton**	**Cross- flow** **Multi – jet – Turgo**	**Water wheel**
Reaction	**-**	**Francis** **Pump- as – turbine**	**Propeller** **Kaplan**	**Propeller** **Kaplan**

Table 6. Groups of water Turbines

The rotating part (called 'runner') of a reaction turbine is completely submerged in water and is enclosed in a pressure casing. The runner blades are designed in a manner such that the pressure difference across their surface imposes lift forces (similar to the principle used for airplane wings) which cause the runner to turn/rotate.

The impulse turbine (as the name suggests) on the other hand is never immersed in water but operates in air, driven by a jet (or jets) of water striking its blades. The nozzle of the penstock converts the head of the water (from forebay tank) into a high speed jet that hits the turbine runner blades that deflect the jet so as to utilize the change of momentum of the water and converting this as the force on the blades – enabling it to rotate.

Impulse turbines are usually cheaper than reaction turbines because there is no need for a pressure casing nor for carefully engineered clearances, but they are also only suitable for relatively higher heads.

5.3.1. Impulse turbines

Impulse turbines are more widely used for micro-hydro applications as compared to reaction turbines because they have several advantages such as simple design (no pressure seals around the shaft and better access to working parts - easier to fabricate and maintain), greater tolerance towards sand and other particles in the water, and better part-flow efficiencies. The impulse turbines are not suitable for low head sites as they have lower specific speeds and to couple it to a standard alternator, the speed would have to be increased to a great extent. The multi-jet Pelton, crossflow and Turgo turbines are suitable for medium heads.

5.3.1.1. Pelton turbine

A Pelton turbine consists of a set of specially shaped buckets mounted on a periphery of a circular disc. It is turned by forced jets of water which are discharged from one or more nozzles and impinge on the buckets. The resulting impulse spins the turbine runner, imparting energy to the turbine shaft. The buckets are split into two halves so that the central area does not act as a dead spot incapable of deflecting water away from the oncoming jet.

Figure 6. Pelton Turbine

The cutaway on the lower lip allows the following bucket to move further before cutting off the jet propelling the bucket ahead of it and also permits a smoother entrance of the bucket into the jet. The Pelton bucket is designed to deflect the jet through 165 degrees which is the maximum angle possible without the return jet interfering with the following bucket for the oncoming jet. They are used only for sites with high heads ranging from 60 m to more than 1000 m.

5.3.1.2. Turgo impulse turbines

The Turgo turbine is an impulse turbine designed for medium head applications. These turbines achieve operational efficiencies of up to 87%. Developed in 1919 by Gilkes as a modification of the Pelton wheel, the Turgo has certain advantages over Francis and Pelton designs for some applications. Firstly, the runner is less expensive to make than a Pelton wheel while it does not need an airtight housing like the Francis turbines. Finally the Turgo has higher specific speeds and at the same time can handle greater quantum of flows than a Pelton wheel of the similar diameter, leading to reduced generator and installation cost. Turgo turbines operate in a head range where the Francis and Pelton overlap. Turgo installations are usually preferred for small hydro schemes where low cost is very important.

Turgo turbine is an impulse turbine where water does not change pressure but changes direction as it moves through the turbine blades. The water's potential energy is converted to kinetic energy with a penstock and nozzle. The high speed water jet is then directed on the turbine blades which deflect and reverse the flow and the water exits with very little energy. Like all turbines with nozzles, blockage by debris must be prevented for effective operation. A Turgo runner looks like a Pelton runner split in half. For the same power, the Turgo runner is one half the diameter of Pelton runner, and so twice the specific speed. e Turgo can handle a greater water flow than Pelton because exiting water doesn't interfere with adjacent buckets.

The specific speed of Turgo runners is between the Francis and Pelton. Single or multiple nozzles can be used. Increasing the number of jets increases the specific speed of the runner by the square root of the number of jets i.e., four jets yield twice the specific speed of one jet on the same turbine.

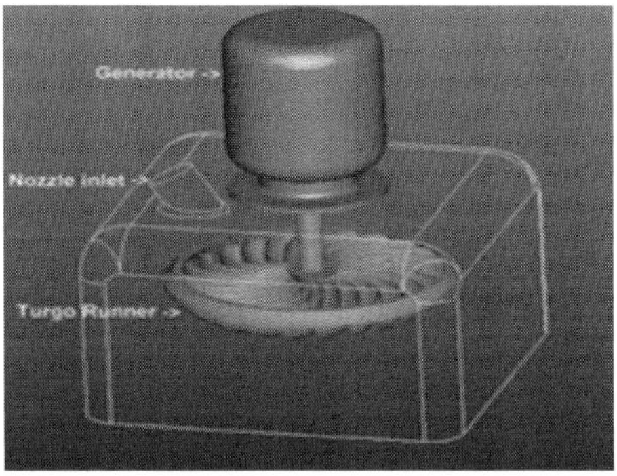

Figure 7. Turgo Turbine

5.3.1.3. Crossflow turbine

Also called a Michell-Banki turbine a crossflow turbine has a drum-shaped runner consisting of two parallel discs connected together near their rims by a series of curved blades. A crossflow turbine always has its runner shaft horizontal (unlike Pelton and Turgo turbines which can have either horizontal or vertical shaft orientation). Unlike most water turbines, which have axial or radial flows, in a crossflow turbine the water passes through the turbine transversely, or across the turbine blades. As with a waterwheel, water enters at the turbine's edge. After passing the runner, it leaves on the opposite side. Going through the runner twice provides additional efficiency. When the water leaves the runner, it also helps clean the runner of small debris and pollution. The cross-flow turbines generally operate at low speeds. Crossflow turbines are also often constructed as two turbines of different capacity that share the same shaft. The turbine wheels are the same diameter, but different lengths to handle different volumes at the same pressure. The subdivided wheels are usually built with volumes in ratios of 1:2. The subdivided regulating unit (the guide vane system in the turbine's upstream section) provides flexible operation, with ⅓, ⅔ or 100% output, depending on the flow. Low operating costs are obtained with the turbine's relatively simple construction. The water flows through the blade channels in two directions: outside to inside, and inside to outside. Most turbines are run with two jets, arranged so that the two water jets in the runner will not affect each other. It is, however, essential that the turbine, head and turbine speed are harmonized. The turbine consists of a cylindrical water wheel or runner with a horizontal shaft, composed of numerous blades (up to 37), arranged radially and tangentially. The edge of the blades are sharpened to reduce resistance to the flow of water. A blade is made in a part-circular cross section (pipe cut over its whole length). The ends of the blades are welded to disks to form a cage like a hamster cage and are sometimes called "squirrel cage turbines"; instead of the bars, the turbine has trough-shaped steel blades.

Figure 8. Crossflow Turbine

The water flows first from the outside of the turbine to its inside. The regulating unit, shaped like a vane or tongue, varies the cross-section of the flow. These divide and direct the flow so that the water enters the runner smoothly for any width of opening. The guide vanes should seal to the edges of the turbine casing so that when the water is low, they can shut off the water supply. The guide vanes therefore act as the valves between the penstock and turbine. The water jet is directed towards the cylindrical runner by a fixed nozzle. The water enters the runner at an angle of about 45 degrees, transmitting some of the water's kinetic energy to the active cylindrical blades. The turbine geometry (nozzle-runner-shaft) assures that the water jet is effective. The water acts on the runner twice, but most of the power is transferred on the first pass, when the water enters the runner. Only ⅓ of the power is transferred to the runner when the water is leaving the turbine.

The crossflow turbine is of the impulse type, so the pressure remains constant at the runner. The peak efficiency of a crossflow turbine is somewhat less than a Kaplan, Francis or Pelton turbine. However, the crossflow turbine has a flat efficiency curve under varying load. With a split runner and turbine chamber, the turbine maintains its efficiency while the flow and load vary from 1/6th to the maximum. The crossflow turbines are mostly used in mini and micro hydropower units less than 2 MW and with heads less than 200 m, since it has a low price and good regulation. Particularly with small run-of-the-river schemes, the flat efficiency curve yields better performance than other turbine systems, as flow in small streams varies seasonally. The efficiency of a turbine is determined whether electricity is produced during the periods when rivers have low heads. Due to its better performance even at partial loads, the crossflow turbine is well-suited to stand-alone electricity generation. It is simple in construction and that makes it easier to repair and maintain than other turbine types. Another advantage is that the crossflow turbines gets cleaned as the water leaves the runner (small sand particles, grass, leaves, etc. get washed away), preventing losses. So although the turbine's efficiency is somewhat lower, it is more reliable than other types. Other turbine types get clogged easily, and consequently face power losses despite higher nominal efficiencies.

5.3.2. *Reaction turbines*

The more popular reaction turbines are the Francis turbine and the propeller turbine. Kaplan turbine is a unique design of the propeller turbine. Given the same head and flow conditions, reaction turbines rotate faster than impulse turbines. This high specific speed makes it possible for a reaction turbine to be coupled directly to an alternator without requiring a speed-increasing drive system. This specific feature enables simplicity (less maintenance) and cost savings in the hydro scheme. The Francis turbine is suitable for medium heads, while the propeller is more suitable for low heads.

The reaction turbines require more sophisticated fabrication than impulse turbines because they involve the use of larger and more intricately profiled blades together with carefully profiled casings. The higher costs are often offset by high efficiency and the advantages of high running speeds at low heads from relatively compact machines. Expertise and precision required during fabrication make these turbines less attractive for use in micro-hydro in developing countries. Most reaction turbines tend to have poor part-flow efficiency characteristics

5.3.2.1. *Francis turbine*

The Francis turbine is a reaction turbine where water changes pressure as it moves through the turbine, transferring its energy. A watertight casement is needed to contain the water flow. Generally such turbines are suitable for sites such as dams where they are located between the high pressure water source and the low pressure water exit. The inlet of a Francis turbine is spiral shaped. Guide vanes direct the water tangentially to the turbine runner. This radial flow acts on the runner's vanes, causing the runner to spin. The guide vanes (or wicket gate) are adjustable to allow efficient turbine operation for a wide range of flow conditions. As the water moves through the runner, it's spinning radius decreases, further delivering pressure acting on the runner. This, in addition to the pressure within the water, is the basic principle on which the Francis turbine operates. While exiting the turbine, water acts on cup shaped runner buckets leaving without any turbulence or swirl and hence almost all of the kinetic or potential energy is transferred. The turbine's exit tube is shaped to help decelerate the water flow and recover the pressure. Francis Turbine and generator Guide vanes at minimum flow setting (cut-away view) Guide vanes at full flow setting(cut-away view) Francis turbines can be designed for a wide range of heads and flows and along with their high efficiency makes them one of the most widely used turbines in the world. Large Francis turbines are usually designed specifically for each site so as to gain highest levels of efficiencies (these are typically in the range of over 90%). Francis turbines cover a wide range of head – from 20 meters to 700 meters, and can be designed for outputs power ranging from just a few kilowatts to one Gigawatt.

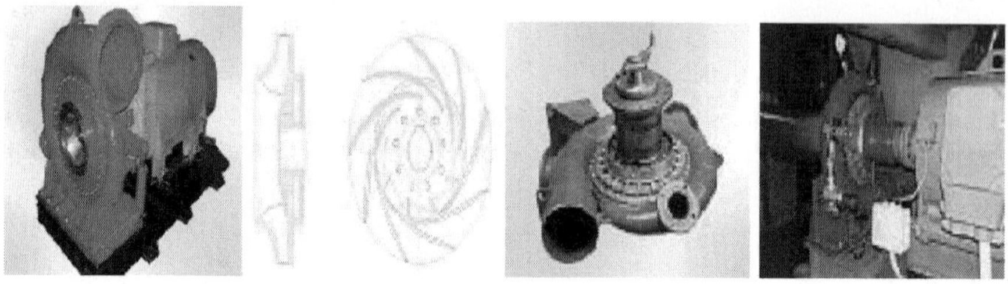

Figure 9. Francis Turbine

5.3.2.2. Kaplan turbine

The Kaplan turbine has adjustable blades and was developed on the basic platform (design principles) of the Francis turbine by the Viktor Kaplan in 1913. The main advantage of Kaplan turbines is its ability to work in low head sites which was not possible with Francis turbines. Kaplan turbines are widely used in high-flow, low-head power production.

The Kaplan turbine is an inward flow reaction turbine, which means that the working fluid changes pressure as it moves through the turbine and gives up its energy. The design combines radial and axial features. The inlet is a scroll-shaped tube that wraps around the turbine's wicket gate. Water is directed tangentially through the wicket gate and spirals on to a propeller shaped runner, causing it to spin. The outlet is a specially shaped draft tube that helps decelerate the water and recover kinetic energy.

The turbine does not need to be at the lowest point of water flow, as long as the draft tube remains full of water. A higher turbine location, however, increases the suction that is imparted on the turbine blades by the draft tube that may lead to cavitations due to the pressure drop. Typically the efficiencies achieved for Kaplan turbine are over 90%, mainly due to the variable geometry of wicket gate and turbine blades. This efficiency however may be lower for very low head applications. Since the propeller blades are rotated by high-pressure hydraulic oil, a critical design element of Kaplan turbine is to maintain a positive seal to prevent leakage of oil into the waterway. Kaplan turbines are widely used throughout the world for electrical power production. They are especially suited for the low head hydro and high flow conditions – mostly in canal based hydro power sites. Inexpensive micro turbines can be manufactured for specific site conditions (e.g. for head as low one meter). Large Kaplan turbines are individually designed for each site to operate at the highest possible efficiency, typically over 90%. They are very expensive to design, manufacture and install, but operate for decades.

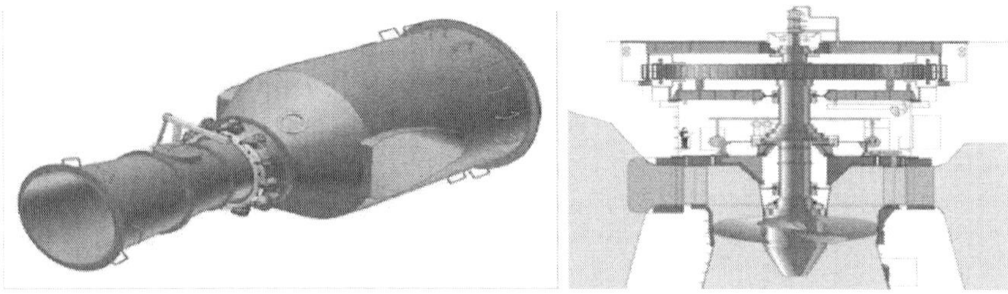

Figure 10. Kaplan Turbine

5.3.3. Turbine selection

Selection of an appropriate turbine to a large extent is dependent upon the available water head and to a lesser extent on the available flow rate. In general, impulse turbines are used for high head sites, and reaction turbines are used for low head sites. Kaplan turbines with adjustable blade pitch are suitable for wide ranges of flow or head conditions, since their peak efficiency can be achieved over a wide range of flow conditions. Small turbines (less than 10 MW) may have horizontal shafts, and even fairly large bulb-type turbines up to 100 MW or so may be horizontal. Very large Francis and Kaplan machines usually have vertical shafts because this makes best use of the available head, and makes installation of a generator more economical. Pelton turbines may be installed either vertically or horizontally.

Some impulse turbines use multiple water jets per runner to increase specific speed and balance shaft thrust. Turbine type, dimensions and design are basically governed by the following criteria:

- Net head
- Variation of flow discharge through the turbine
- Rotational speed
- Cavitation problems (quality of water available from penstock)
- Cost

The main criterion considered in turbine selection is the net head. The figure given above (Turbine Application Chart) specifies the range of operating heads for each turbine type. The figure above and the table below show some overlapping, so that for a given head several types of turbines can be used. The selection is particularly critical in low-head schemes, where large discharges need to be handled to be economically viable.

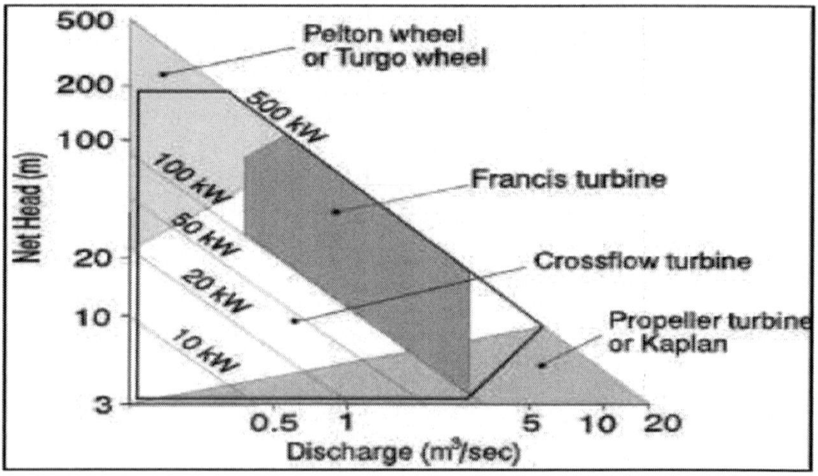

Figure 11. Turbine Application Chart based on Head and Discharge

Turbine Type	Typical range of heads (H = head in m)
Hydraulic wheel turbine	$0.2 < H < 4$
Archimedes' screw turbine	$1 < H < 10$
Kaplan & Propeller	$2 < H < 40$
Francis	$10 < H < 350$
Pelton	$50 < H < 1300$
Michell-Banki	$3 < H < 250$
Turgo	$50 < H < 250$

Table 7. The selection of turbine according to the head

5.3.4. *Turbine efficiency*

A significant factor in the comparison of different turbine types is their relative efficiencies both at their design point and at reduced flows. Typical efficiency curves are shown in the figure below. An important point to note is that the Pelton and Kaplan turbines retain very high efficiencies when running below design flow; in contrast the efficiency of the Crossflow and Francis turbines falls away more sharply if run at below half their normal flow. Most fixed-pitch propeller turbines perform poorly except above 80% of full flow.

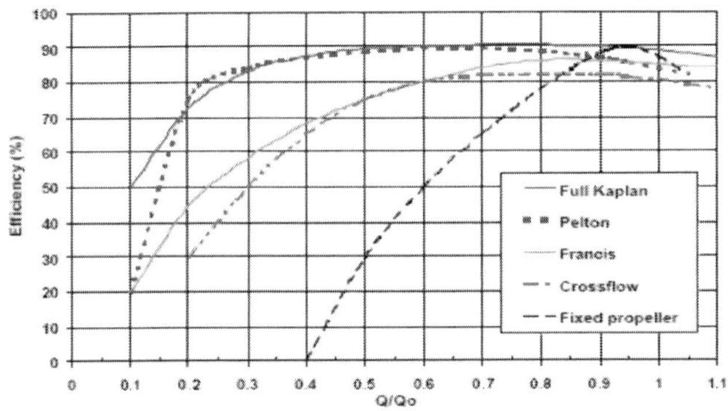

Figure 12. Efficiency of Various Turbines based on Discharge rate

6. Hydro power in Iraq

In Iraq there are two big rivers (Tigris& Euphrates) where a many dams are built, that gives the ability to establish a hydro electric power plants on these dams such as (mousal , hadeetha , samara , diala ,dokhan……etc) , which are distributed on the map below.

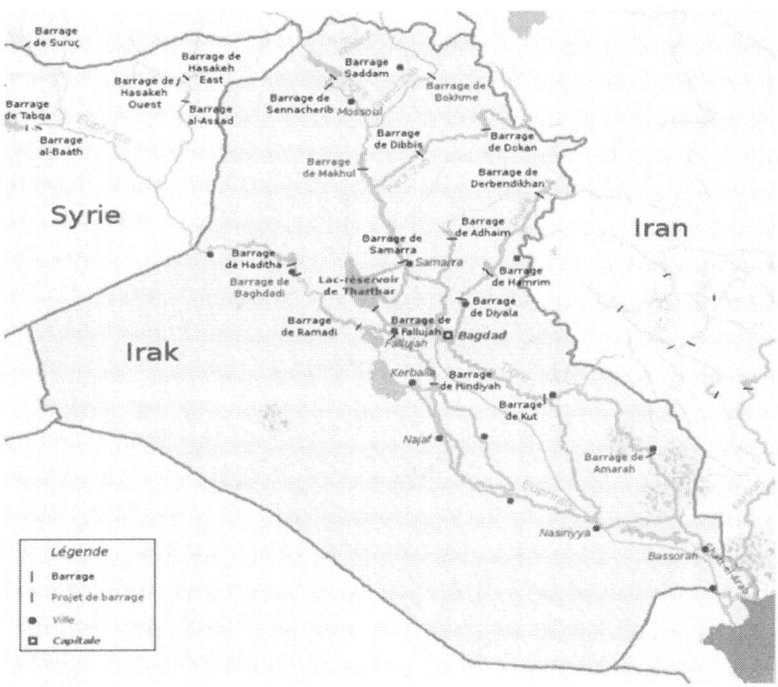

Figure 13. The distribution map of Dams in Iraq

Most of these dams have been used in the generation of electricity by building a hydro electric power plants on it and the output power of these plant in Iraq are shown in table 6 below. The total generated power in Iraq is about **10171.5 MW** as we see from this table the hydro power generated is about **2489** Mw which represent a (25%) the amount of hydroelectric power generated in Iraq varies according to the incoming quantity of water that enter from these rivers.

Name of hydro power plant	*Generated power output (Mw)*
Mosul	1050
Hadeetha	60
Hindia	15
Dukhan	400
Al- Kufa	5
Hmreen Dam	50
Samaraa	84
Al- Qadssia	660
Darbandikhan	240

Table 8. Hydro power plants in Iraq

6.1. Hydropower transportation and power conservation

In order to achieve the maximum (optimum) utilization of water and hydropower of the rivers in Iraq we study and calculate the elevation of water at the dams in the whole country such as (Hadeetha, Samara, Diala, Hindia, Kufa and Abasia ….etc) and the elevation of the middle and south province center (Baghdad, Basra, Babylon, Kut, Nasria, Missan, and Samawa), by using the Google Earth program and then calculate the difference in elevation. Figure (14) shows the map of the region of study.

Since the incoming water to Iraq was decrease last years and the high demand on the water and electricity, the concentration of study was on the water supply and the amount of electricity that can be generated from this quantity of water. In addition to the lack of fresh water that suitable for the everyday using because the high ratio of salts in the water that arrive to some cities through Tigris& Euphrates rivers, the problem that the other will be faced with the continuous reduction in the arrived quantity of water to Iraq, and the increasing in pollution levels of these rivers in the future. Where the expected pollution levels will cross over the permition level at which the processing of water become more difficult that the traditional water station can reach it.

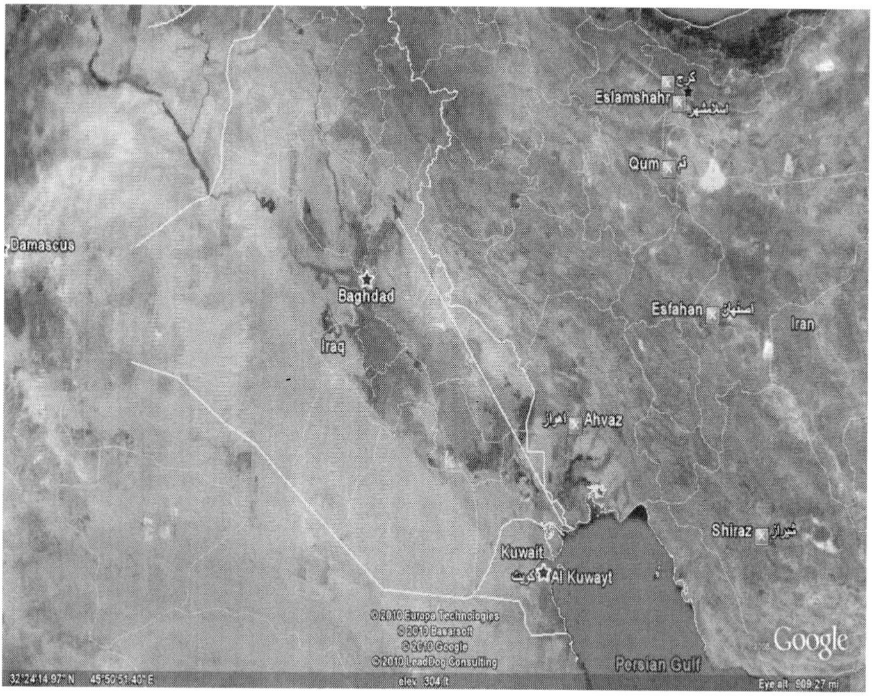

Figure 14. The map of the region of study

static (at no flow) head loss = 80 ft				dynamic(at design flow) head loss =100 ft		
Q	D	Po/p	EPM	D	Po/p	EPM
1000	30	28.8	20736	28	36	25920
2000	39	57.6	41472	37	72	51840
3000	45	86.4	62208	43	107	77760
4000	50	115.2	82944	48	144	103680
5000	55	144	103680	52	180	129600
6000	59	172.8	124416	56	216	155520
7000	62	201.6	145152	59	252	181440
8000	65	230.4	165888	62	288	207360
9000	68	259.2	186624	65	324	233280
10000	71	288	207360	68	360	259200

Head = 400ft , Distance = 466 mile, Sys. Efficiency =50%, Q = flow rate(in gpm), D = pipe line diameter(in inch), Po/p = power output(in KW), EPM = Energy per month(in KWH)

Table 9. Shows the Relation Between The Q (flow rate) and the o/p Power and Pipeline Diameter and The Other Parameters are Constant.

From the results shown in the above tables we see that. The difference in elevation(head) between Basra city and Hadeetha dam is 400 ft and the distance is 466 mile approximately. To supply the city by 1000 gpm we need a pipeline with 28 inch diameter or more, and to supply the city by 10000 gpm the pipeline of 68 inch diameter or more can be used. These pipeline can produce a hydroelectric power from 36 KW to 360 KW which can save an energy from 25920 KWH to 259200 KWH per month, in addition to the power saving, through the way of supplying water with high pressure by using a pipeline, which will consumed in the case of lifting that quantities of water from the river (low lift) at the destination point to the water station or for any other using of water.

The difference in elevation(head) between Basra city and Diala (Hemrin) dam is 290 ft and the distance is 341 mile approximately. To supply the city by 1000 gpm we need a pipeline with 28 inch diameter or more, and to supply the city by 10000 gpm the pipeline of 68 inch diameter or more can be used. These pipelines can produce a hydroelectric power from 26.1 KW to 261 KW , which can save an energy from 18792 KWH to 187920 KWH per month in.

The difference in elevation(head) between Basra city and Abasia dam is 70 ft and the distance is 260 mile approximately. To supply the city by 1000 gpm we need a pipeline with 36 inch diameter or more, and to supply the city by 10000 gpm the pipeline of 86 inch diameter or more can be used. These pipelines can produce a hydroelectric power from 6.3 KW to 63 KW , which can save an energy from 4536 KWH to 45360 KWH per month

Figure (15) below shows the relations between the flow rate in gpm with pipeline diameter in inch that represent the quantity of water supplied to the Basra city from the dams (Hadeetha, Diala, Hindia, Abasia). The one who see this relations utilize that the graph of Hadeetha dam and Diala is complying in spite of the difference in the dams head, because the difference in the distances, where long distance causes high friction loss ,so that to decrease this friction loss the pipeline diameter must be large, and the same thing shown for Abasia and Hindia dams.

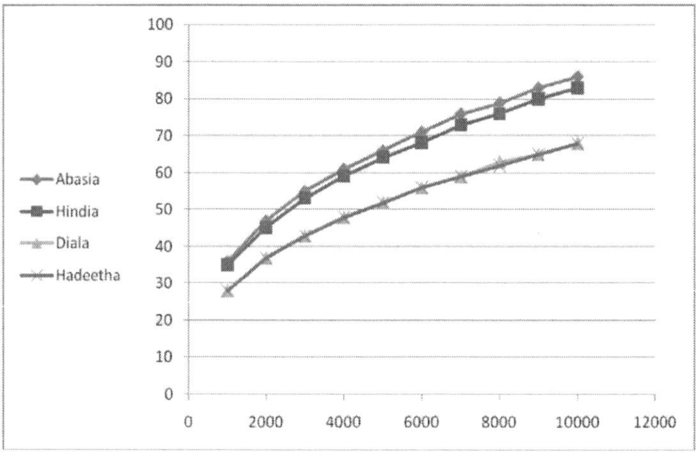

Figure 15. The Relation Between Flow Rate (Q) in gpm and Pipeline Diameter in inch For Supplying Water to Basra City from Dams

Figure (16) below shows the relations between the flow rate in gpm that represent the quantity of water supplied to the Basra city from the dams (Hadeetha, Diala, Hindia, Abasia) with the hydroelectric generated power in KW. The idea behind this graph is that the slopes of the lines increase with the increasing of head at the dams, so that the hydro generated power from Hadeetha dam is the higher and the lower power from Abasia dam for the same quantity of the flow rate (Q) gpm.

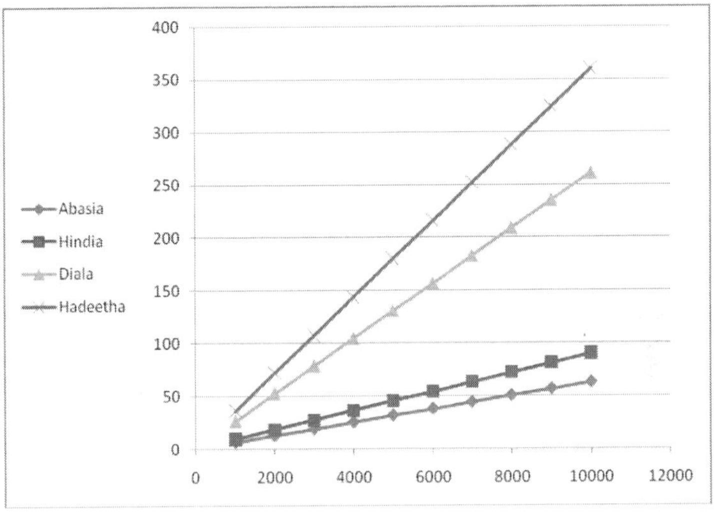

Figure 16. The Relation Between Flow Rate (Q) in gpm and The Hydroelectric Generated power in KW From Supplying Water to Basra City from Dams

From all the above , if the one need to adopted this study to execute or establishing a project of water pipelines and a micro & mini hydroelectric generators, which need to a further studying about cost calculations and the obstacles or difficulties that may be faced him, because this is a theoretical study and the practical project will require a solution to a practical problems that the pipe lines intersect it , like mount , river , villages , ...etc.

6.2. Hydro power transportation

The use of pipeline in water transportation guarantee three type of power conservation the first is hydroelectric generation by installing a turbine in the way of this pipeline, the second type of power conservation when it (pipeline) may also being used instead of low lift pumping station for the water treatment plant in this operation the power can be conserved approximately 30% -40% of the whole total power consumed in the water treatment plants in Babylon province when the difference in elevation between the river and the reservoir tank is about 5-10 m (head). The using of pipeline in the supply of water directly conserved power because it doesn't need to convert the hydropower (head & flow rate) into mechanical power and then into electrical power that used in the low lift part of water treatment plant that need to convert the electrical power into mechanical power to lift the water from the river to the reservoir tank, because of the efficiency of conversion is always less than 100% (turbine and generator efficiency), this operation leads to conserve the power approximately with 50% or more of the hydropower, which may be consider the third type of power conservation. This type of power conservation is very clear in the water treatment plant at Bekal waterfall figure (17) when the intake established on it, but penstock(pipeline) length is tenths meter to the reservoir tank the length that may extended to kilometer or handered of kilometer in the other place, which represent the low lift in the traditional water treatment plant. Also figure 18 shows the water pump used for pumping water in the water treatment station.

Figure 17. Shows the intake and the penstock (pipeline) of Bekal water treatment plant

Figure 18. Shows the high head pump station

7. Summary

Water is one of our most valuable resources, and hydropower makes use of this renewable treasure. *Hydropower* traditionally represents the energy generated by damming a river and using turbine systems to generate electrical power. In the ancient times waterwheels were used extensively, but it was only at the beginning of the 19th Century with the invention of the hydro turbines that the use of hydropower got popularized. Small-scale hydropower was the most common way of electricity generating in the early 20th century. Hydropower will continue playing an important role throughout the 21st Century, in the world of electricity supply. Hydropower development does have some challenges besides the technical, economic and environmental advantages, which shares with the other power generation technologies. At the beginning of the new Millennium hydropower provided almost 20% (2600 TWh/year) of the electricity world consumption (12900 TWh/year).

The countries have different criteria in the classification of hydro power plants, the hydro plants may be also classified according to the "Head" or the vertical distance through which the water is made to impact the turbines.

Power generation from water depends upon a combination of head and flow. Both must be available to produce electricity. Water is diverted from a stream into a pipeline, where it is directed downhill and through the turbine (flow). The vertical drop (head) creates pressure at the bottom end of the pipeline. The pressurized water emerging from the end of the pipe creates the force that drives the turbine. The turbine in turn drives the generator where electrical power is produced. More flow or more head produces more electricity. Electrical power output will always be slightly less than water power input due to turbine and system inefficiencies. Water pressure or Head is created by the difference in elevation between the water intake and the turbine.

The theoretical power (P) available from a given head of water is in exact proportion to the head and the quantity of water available.

$$P = Q \times H \times e \times 9.81 \left(kW \right)$$

Where

P Power at the generator terminal, in kilowatts (kW)

H The gross head from the pipeline intake to the tail water in meters (m)

Q Flow in pipeline, in cubic meters per second (m^3/s)

e The efficiency of the plant, considering head loss in the pipeline and the efficiency of the turbine and generator, expressed by a decimal (e.g. 85% efficiency= 0.85)

9.81 is a constant and is the product of the density of water and the acceleration due to gravity (g)

The penstock pipe (pipeline) transports water under pressure from the forebay tank to the turbine, where the potential energy of the water is converted into kinetic energy in order to rotate the turbine. The penstock is often the most expensive item in the project budget – as much as 40 percent is not uncommon in high-head installations. It is therefore worthwhile to optimize its design in order to minimize its cost. Head loss due to friction in the penstock pipe depends principally on the velocity of the water, the roughness of the pipe wall and the length and diameter of the pipe. The losses decrease substantially with increased pipe diameter. Conversely, pipe costs increase steeply with diameter. Therefore, a compromise between cost and performance is required. Several factors should be considered when deciding which material to use for a particular penstock: design pressure, the roughness of the pipe's interior surface, method of joining, weight and ease of installation, accessibility to the site, design life and maintenance, weather conditions, availability, relative cost and likelihood of structural damage. The pressure rating of the penstock is critical because the pipe wall must be thick enough to withstand the maximum water pressure; otherwise there will be a risk of bursting. The penstock becomes more expensive as the pressure rating increases. The most commonly used materials for a penstock are HDPE, uPVC and mild steel because of their suitability, availability and affordability. Layout of the penstock pipelines depends on their material, the nature of the terrain and environmental considerations; they are generally surface-mounted or buried underground.

The hydrostatic pressure created from the head must be determined so that a suitable wall thickness can be determined. This pressure is given by Equation, **Pressure = p*g*H**

Turbine is the main piece of equipment in the hydro power scheme that converts energy of the falling water into the rotating shaft power. The selection of the most suitable turbine for any particular hydro site depends mainly on two of the site characteristics – head and flow available. All turbines have a power-speed characteristic. This means they will operate most efficiently at a particular speed, head and flow combination. Thus the desired running speed of the generator or the devices being connected/ loading on to the turbine also influence selection. Other important consideration is whether the turbine is expected to generate power at part-flow conditions. The design speed of a turbine is largely determined by the head under which it operates. Turbines can be classified as high head, medium head or low head machines. They are also typified by the operating principle and can be either impulse or reaction turbines.

In order to achieve the maximum (optimum) utilization of water and hydropower. The use of pipeline in water transportation guarantee three type of power conservation the first is hydroelectric generation by installing a turbine in the way of this pipeline, the second type of power conservation when it (pipeline) may also being used instead of low lift pumping station for the water treatment plant and the using of pipeline in the supply of water directly conserved power because it doesn't need to convert the hydropower (head & flow rate) into mechanical power and then into electrical power which used in the low lift part of water treatment plant that need to convert the electrical power into mechanical power to lift the water from the river to the reservoir tank, because of the efficiency of conversion.

Author details

Mohammed Taih Gatte and Rasim Azeez Kadhim
Ministry of Sciences and Technology, Babylon Department, Hilla, Iraq

8. References

[1] Energy Producing Systems ,Hydro Power, Energy For Missouri: Today and Tomorrow - *Educator's Guide*

[2] Dilip Singh, " Micro Hydro Power Resource Assessment Handbook ", APCTT, September 2009.

[3] Micro -Hydro Power , Practical Action, The Schumacher Centre for Technology and Development, Bourton on Dunsmore, Rugby, Warwickshire, www.practicalaction.org

[4] Micro-Hydropower Systems: A Buyer's Guide, Her Majesty the Queen in Right of Canada, 2004, ISBN 0-662-35880-5

[5] Jacques Chaurette p. eng. " Tutorial Centrifugal Pump Systems ", copyright 2005, Web site: www.fluidedesign.com

[6] Md Tanbhir. Hoq, Nawshad U. A., Md. N. Islam, IbneaSina ,Md. K. Syfullah, Raiyan Rahman, " Micro Hydro Power: Promising Solution for Off-grid Renewable Energy Source ", International Journal of Scientific & Engineering Research, Volume 2, Issue 12, December-2011 1 ISSN 2229-5518 IJSER © 2011 http://www .ijser.org

[7] Leo Lovel, Scott Craig, S. Cody Maher, Jesse Ross, Brian Van Stra, MicroHydro Turbine A Feasibility Study

[8] Minister of Natural Resources Canada 2001 – 2004. " Small Hydro Project Analysis "

[9] Mohammed Taih Gatte, Rasim Azeez Kadhim, Farhan Leftah Rasheed," Using Water Energy for Electrical Energy Conservation by Building of Micro hydroelectric Generators on The Water Pipelines That Depend on The Difference in Elevation", IEEE 2011.

[10] Yunus A. Cengel, Michael A. Boles. " Thermodynamics an Engineering Approach ". Fifth Edition

[11] Brian B. Yanity, "Cold Climate Problems of a MicroHydroelectric Development on Crow Creek", The Arctic Energy Summit, Anchorage, Alaska 2007 – 2008.

Web sites :

http://web.archive.org/web/20070509191717/http://www.microhydropower.net/intro.html

http://www.bbgreeneurope.eu/vodne-turbiny/snekova-turbina

http://www.nooutage.com

Chapter 5

Earth Shelters; A Review of Energy Conservation Properties in Earth Sheltered Housing

Akubue Jideofor Anselm

Additional information is available at the end of the chapter

http://dx.doi.org/10.5772/51873

1. Introduction

Earth sheltering is an age long traditional practice. In modern times its benefits has prompted new definitions for its practice. With the potential thermal conservation qualities and physical characteristics of earth as a building mass, earth shelters can now be defined as structures built with the use of earth mass against building walls as external thermal mass, which reduces heat loss and maintains a steady indoor air temperature throughout the seasons. The popularity of earth sheltering was advanced mostly by research in energy conservation in residential housing. Originally conceived as dwellings developed by the utilization of caves within the traditional context, its evolution through technologies led to the construction of customized earth dwellings all across the globe. These structures in the past were built by people not schooled in any kind of formal architectural design or with identifiable building techniques rather they depended on the cover the very structure of the earth could provide them for purposes of shelter, warmth and security. Investigations into the traditional earth sheltered dwellings also identified sunken earth houses with characteristics that suggested potentials in passive building insulation which utilizes ground thermal inertia.

In the view of some researchers on earth supported housing, building underground provides energy savings by reducing the yearly heating and cooling loads in comparison with known conventional structures. Not only is the temperature difference between the exterior and interior reduced, but mostly because the building is also protected from the direct solar radiation [1].

One significant value of earth-sheltered housing and the reason for its evaluation is its potential energy savings when compared to conventional aboveground housing. This potential is based on several unique physical characteristics. The first of these characteristics is in the reduction of heat loss due to conduction through the building envelope because of the high density of the earth. According to [2], in an earth sheltered building even at very shallow depths and given normal environmental conditions, the ground temperatures seldom reaches the outdoor air temperatures in the heat of a normal summer day. This condition allows the conducting of less heat into the house due to the reduced temperature differential.

In the case of colder climates, it was noticed that during winters the rate of heat loss in bermed (earth supported) structure was less in comparison to that in on-grade structures. This indicates through results that the floor surface temperature increased by 3° C for a 2.0m deep bermed structure due to lower heat transfer from the building components to the ground, thus suggesting the presence of passive heat supply from the ground even at the extreme cold temperatures of winter [3]. This evidently contributes as a factor for energy saving in earth shelter buildings in cold climates.

Other characteristics include the reduction of air infiltration within the dwelling which is mainly surrounded by earth walls with very little surface area exposed to the outside air. These characteristics have been investigated in previous studies and the analysis on each location provides results and findings in terms of climatic effects, design styles and residential activities of the dwellers that bring about the unique energy saving value of these buildings.

Single unit earth sheltered houses are unique energy conservation ideas based on their earth contact characteristics as mentioned above. In order to achieve the maximum benefits from earth sheltered housing, its application could be examined also at an entirely community scale rather than simply at the scale of individual houses. One of the biggest challenges to the overall performance of earth sheltered housing would be the built conventional surroundings. While contemporary use of earth sheltering is confined to individual homes built on single plots of land or a small cluster of houses which will absolutely be affected by the surrounding conventional structures around, the traditional use encompassed entire communal design or villages that will stay within the same conditions the micro-environment provides. This communal development option is identified to be most effective as isolated pockets of earth sheltered houses do not really reach the scale needed for sustainable development [4]. Earth sheltered mass-housing may thence become the general concept for design and building with earth whereby entire communities are created, enjoying dual land use by locating all housing underground [5]. If a single case of earth sheltering is found to have significant advantages, these advantages can only increase in magnitude if applied to whole communities.

2. Fundamentals of Earth sheltered housing

The values of energy conservation in earth shelters are dependent on certain principles. These principles which form the ground rules for the design and construction of earth sheltered dwellings have been existent since prehistoric periods. Earth sheltered homes were primarily developed for shelter, warmth and security for the earliest human dwellers.

Most of the recorded cases of these shelters are found extensively in areas like Asia and Northern Africa. In one of the earliest cases in Japan was discovered the oldest human habitation in a layer of earth about 600,000 years old in Kamitakamori, Miyagi Prefecture. Archaeologists from the Tohoku Paleolithic Institute, Tohoku Fukushi University and other institutes believed that the finding may be one of the oldest in the world. There are only a few remains of human dwelling structures from the early Paleolithic period in the world, as early humans such as the Peking-man lived in caves. Researchers believed the dwellings were built by primitive man who appeared some 1.6 million years ago and likely reached Japan 600,000 years ago at the latest, according to the archaeologists. The buildings could have been used as a place to rest, a lookout for hunting, a place to store hunting tools or to conduct religious rites.

In Tunisia, residents of Matmata were discovered to have lived in manmade caves for centuries (Figure 1). Here rooms were carved into the soft rock to create atrium houses that had several excavated rooms with up to 4 to 10 meter high and vaulted ceilings opening out onto a single sunken courtyard. The original objective for going below the ground in this case was to protect the inhabitants from the extremes of daytime North African heat and nighttime cold, typical of this desert region.

Figure 1. Aerial view of a typical Matmata earth shelter dwelling. Image by Tore Kjeilen

However through the years, more modern earth sheltered dwellings were revealed as studies on the earliest forms of human settlements progressed. In China, modern earth shelters habitats were discovered with histories that dated back to before 2000 B.C. This type of habitats were commonly called cave dwellings as they were strictly home units hewed out of the mountains. It is believed that underground housing preceded above ground housing in this area. Studies on these existing Chinese earth habitats presented analytical data on the climatic and topographical relationships to the unique design elements utilized to attain living comfort by the cave shelter dwellers. Such analysis as the rain, wind, sun and seasonal weather conditions that exist in these areas where these dwellings were located possibly necessitated the advantage of its existence in these locations [6]. Analysis on each location also provided results and findings in terms of climatic effects, design styles and residential activities of the dwellers. In the North-west of China, variety of these structures evolved, ranging from the cave dwelling units to the more advanced subterranean types. In the case of the traditional subterranean homes in China (called '*yao dong*'), rooms were dug into loose, silty soil to primarily combat the hot summers and bitterly cold winters. In the early 20th century the provinces of Shanxi, Jiansu and Henan still had traditional dwellers that faced with the need to preserve agricultural land and housing for their people, dug entire cities beneath their lands. Today, it is still believed that more than 10 million Chinese live underground, perhaps the largest number of troglodytes ever to inhabit a single region. The Shanxi homes (Figure 2, 3 and 4) were buried at depths of up to 10 meters with their underground homes built around courtyards. This atrium-style design offer ample sunlight as well as surface spaces for other activities.

Research conducted in [6] also provided analytical data on climatic and topographical relationships to the structural design styles with single unit design solution, multi unit designs and finally urban planning initiatives on how to achieve a sunken city that exists beneath rather than above ground level as seen in Figure 2 below. Also fascinating in discovery included methods and techniques of ventilating the building units naturally. Such natural ventilation techniques are viewed today as ideas that advanced the notion of passive aeration of interiors which ultimately is a cost and energy efficient alternative to the whole process of earth sheltered housing.

Figure 2. Aerial view of an earth shelter neighborhood in *Lian Jiazhuang*, Shanxi Province, North-western China

(a) (b)

Figure 3. (a) Courtyard view of an Atrium type subterranean earth shelter dwelling in *Lian Jiazhuang*, Shanxi Province. (b) Interior view of a typical room space. Image by Kevin Poh.

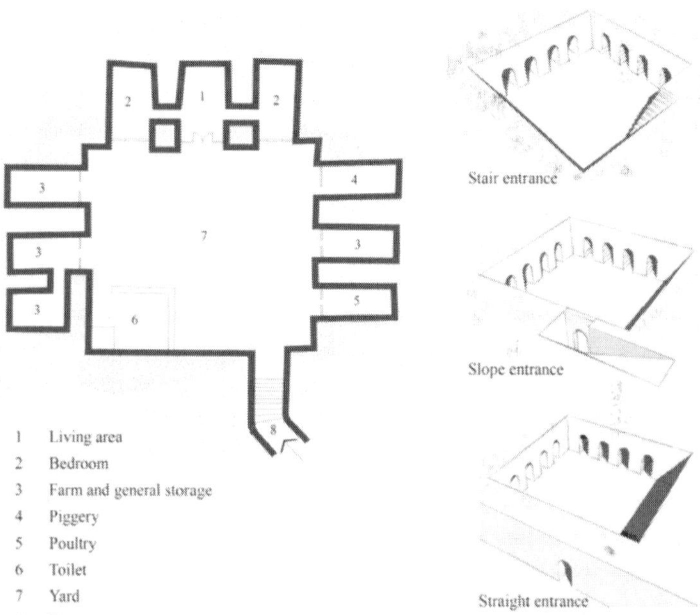

Figure 4. A typical earth shelter home layout in North-western China

With the challenges of global warming and fossil energy reduction, energy saving ideas has become an essential element in building designs and occupation. Since energy conservation is the practice of saving energy use without compromising occupant thermal comfort [7], building below the ground thence presents certain fundamentals that with the aid of research can significantly influence energy conservation efforts in modern housing. From reviews of the basic background of traditional earth sheltered housing, the fundamental objectives for building below the ground and significant energy conservation principles are listed as follows:

1. Indoor temperature enhancement based on the natural principles of annual heat storage (PAHS) whereby the earth collects free solar heat all summer and cools passively while heating the earth around it, and keeping warm in winter by retrieving the stored heat from the soil in winters. This dual function presents a scenario that makes the practice of earth sheltered housing effective in both hot and cold climates.
2. Huge temperature differential between the ground temperatures and the outdoor air temperatures. In this case the normal ground temperature seldom reaches the outdoor air temperatures in the heat of a normal hot day, thereby conducting less heat into the house due to the reduced temperature difference.
3. Building protection from the direct solar radiation, thereby elimination the challenge of direct thermal load due to heat radiation through the building envelope.

Apart from the energy values which the subsurface climate of the earth provides, the other significant characters beneficial to earth shelters includes the major goal of recycling surface space by relocating functions to underground, by this earth shelters liberates valuable surface space for other functional uses and improves ground surface visual environment, open surfaces for landscaping and thus a more greener atmosphere.

3. Modern construction techniques and design typology

The structural make up of a typical earth shelter house is made up of the supporting members and the compacted backfills in which case strength and composition can determine the ability to withstand overhead loads of moisture, dead and live loads, the distribution of which depend on the compaction strength of the backfill or supports.

However in modern designs, the supports are the parts of the house that brace against the side walls of soil and overlaying roof members that are made of backfills as in the case of underground homes. The design method and material choice will determine the resistance to failure of these structural members. In the traditional construction scenario where the earth-soil is used as building material; its strength is determined by the soil stability, which goes to improve the resistance to wind and in most cases rain erosion.

3.1. Earth shelter structural integrity

The structural make up of earth homes is mainly made up of the supporting members and the compacted backfills. As earlier mentioned, the strength and composition of the material used as backfill can determine the ability to withstand overhead loads. The supports are the parts of the house that brace against the side walls of soil and overlaying roof members that are made of backfills. The building design method and material choice will determine the resistance to failure of these structural members. In the case where the earth-soil is used as building material, its strength is determined by the soil stability, which goes to improved the resistance to wind and rain erosion. In most earth shelter construction the significant structural areas are the soil, walls and roof area. Apart from serving as a building material, the soil-walls of the shelter trench are regarded as the most valuable structural member of the Earth house structure. It provides the necessary support a normal wall gives in an ordinary house design. Nevertheless, not all soil types are efficient in use for earth sheltered house construction. From studies it is identified that the best soils are granular, such as sand and gravel. These soils compact well for bearing the weight of the construction materials and are very permeable, which means they allow water to drain quickly. The poorest soils are cohesive, like clay, which may expand when wet and has poor permeability. Soil tests, offered through professional testing services, can determine load-bearing capability of soils and possible settlements that may occur after construction. Study in [6] revealed certain traditional considerations for deciding the depth, thickness of mass and curvature of the support ceilings (vault) of the Chinese earth homes which can also be applied in modern day construction of earth shelters (figure 5).

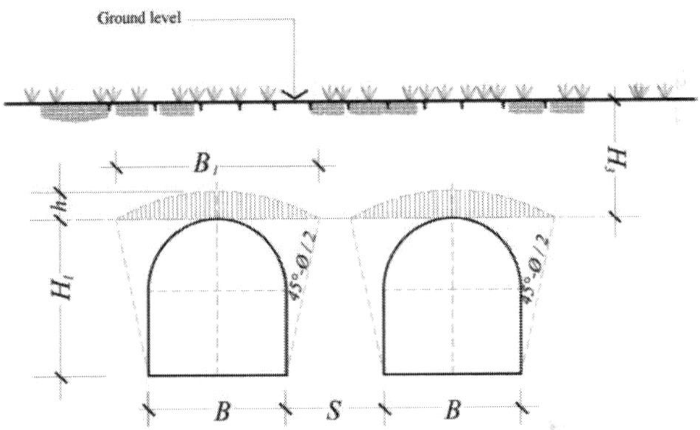

Figure 5. Structural consideration for a typical room space excavation in the Shanxi traditional earth shelters

$h = 1 \sim 2,\ Ø = 18°.$

$\frac{1}{2} B_1 = \frac{1}{2} B + H_1\, tg\,(45° - Ø/2)$

$= 3.5/2 + 3\, tg\,(45° - 18°/2)$

$\frac{1}{2} B_1 = 1.75 + 2.19 = 3.94\ m$

Then S = Thickness of Earth thermal mass wall

H_3 = Extent of depth clearance

Assuming B (room span) = 3.5m and H_1 (room height) = 3m

H_3 = Depth from ground surface to ceiling. This should be greater than h

The Dotted/shaded area indicates possible fault lines due to the pressure from the overlaying earth mass

Varieties of techniques have been used in the past for earth shelter wall construction. The construction materials for the walls of each type of structure will vary, depending on characteristics of the site, climate, soils, and design. However, general guidelines show that houses more deeply buried require stronger, more durable structural walls. Walls must provide a good surface for waterproofing and insulation to withstand the pressure and moisture of the surrounding ground. When soil is wet or frozen, the pressure on the walls and floors increases as pressure also increases with depth.

For the traditional earth supported homes built in the Chinese and Arid (dessert) climatic regions, there usually is no use for supporting walls as the naturally compressed soil structure already serves the function. However through recent research on improving the state of earth homes for most other climatic regions, the walls of Earth homes can be made of various materials ranging from Compressed Earth bricks to Concrete, while providing cavities and drainage patterns to aid damp proofing. In most earth home designs, the roof is usually the most challenging part of the entire structure. With recent ideas in ecology, the roof of earth shelters assume interesting landscaping functions. Especially for earth supported shelters which already posses the natural materials of earthen walls and members, the roof can also be finished to assume a natural finish too. Since the basic idea of this study is to discover techniques to achieve high performance as possible, the basic structural form for constructing the earth shelter roof is as follows:

1. A frame strong enough to support the dead load brought by the soil overlay, rain, snow and ice loads where applicable.
2. A solid deck built over the frame and a waterproof membrane installed on the deck prior to final earth cover.
3. Treated soil backfill placed on the membrane (as the roof layer) and covered with a fine thick layer of soil. The roof will either grow a vegetation of its own or become a life garden depending on the appropriate type of maintenance.

Reinforced concrete is the most commonly used structural material in earth shelter construction. Products like Grancrete and Hycrete are becoming more readily available. They claim to be environmentally friendly and either reduce or eliminate the need for additional waterproofing. However, these are new products and have not been extensively used in earth shelter construction.

Some other unconventional approaches are also utilized in earth shelter construction. These techniques utilize recycled material of various forms and applications. One of such approaches is referred to as an Earth ship (figure 6). These houses are built to be self-contained and independent; their design allows occupants to grow food inside and to maintain their own water and solar electrical systems [8]. Some builders believe they have proven the design's ability to tap into the constant temperature of the earth and store additional energy from the sun in winter. These Earth ships carry out their environmentally conscious theme by employing unusual building materials in the form of recycled automobile tires filled with compacted earth for thermal mass and structure. While the tires form the major structural frames for the building, aluminum or tin cans are used for filling minor walls that are not load-bearing. Foam insulation can be applied to exposed exterior or interior walls and covered with stucco. Interior walls are also dry-walled giving it a conventional look.

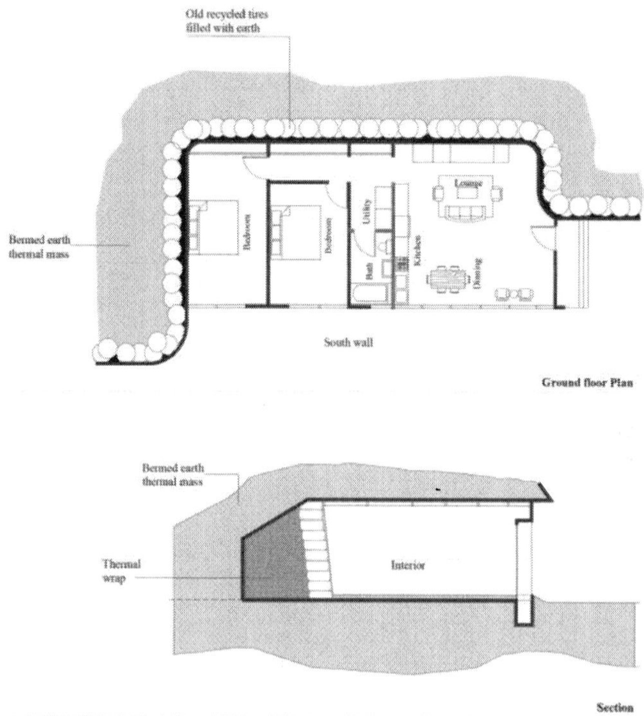

Figure 6. An Earth Ship design, using recycled materials

3.2. Earth shelter construction typology

Earth sheltered houses are often constructed with energy conservation and savings in mind. Though techniques of earth shelter construction have not yet become common knowledge, study into the most efficient application of the earth shelter principles reveals classifications of the major typologies that are utilized in the construction of earth houses. These major construction concepts are the Bermed or banked with earth type and the Envelope or True underground type. The energy conservation values of these typologies also vary depending on climate and physical challenges indigenous to each typology (table 1).

a. **Bermed earth shelter:** In this type of construction, earth is piled up against exterior walls and heaped to incline downwards away from the house. The roof may, or may not be, fully earth covered, and windows/openings may occur on one or more sides of the shelter. Due to the building being above ground, fewer moisture problems are associated with earth berming in comparison to the fully underground construction. Other variations of bermed construction are the elevational and in-hill construction (figure 7). This type of construction is particularly appropriate for colder climates. With regards to energy efficiency in colder climates, all the living spaces may be arranged on the side of the house facing the equator. This provides maximum solar radiation to the most frequently used spaces like bedrooms, living rooms, and kitchen spaces [9]. Rooms that do not require natural daylight and extensive heating such as the bathroom, storage and utility rooms are typically located on the opposite in-hill side of the shelter. The compact configuration of this construction provides it with a greater ratio of earth cover to exposed wall thereby improving its energy performance benefits through the earth-contact principles. However the case for both climates, the three major determinants for the building orientation remains the sun, wind and outside views. Proper orientation with respect to solar path and wind is significant for energy savings.

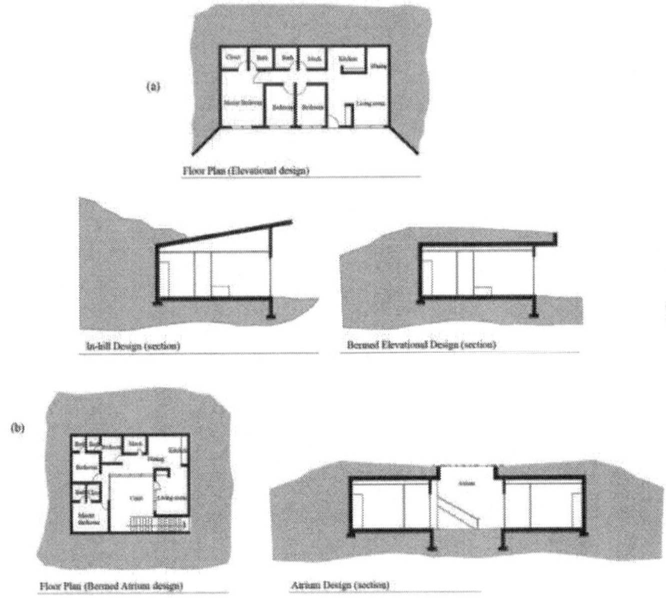

Figure 7. (a) Elevational (beremed) and in-hill designs, (b) Atrium (bermed) design

b. **Envelope or True underground earth shelter:** In the true underground construction, the house is built completely below ground on a flat site, with the major living spaces surrounding a central outdoor courtyard or atrium. The windows and glass doors that are on the exposed walls facing the atrium provide light, solar heat, outside views, and access via a stairway from the ground level. The atrium effect offers the potential for natural ventilation. In the view of some researchers, this concept reduces the energy conservation properties in colder climates mostly due to the reduced solar exposure within the courtyard or atrium opening [9]. However recent studies in the area of soil temperature analysis with respect to energy conservation in earth shelters, provides information on the prospect of efficient underground earth shelter design. Such studies as in [10], provides mathematical method for predicting the long-term annual pattern of soil temperature variations as a function of depth and time for different soils and soil properties that are stable over time and depth. The likes of these studies were utilized by John Hait's [11] in his book on Passive Annual Heat Storage (PAHS) to advance the ideas of earth shelter housing. With the development of modern passive solar building design, during the 1970s and 1980s a number of techniques are developed to enabled thermally and moisture-protected soil to be used as an effective seasonal storage medium for space heating, with direct conduction as the heat return method. Other variations of the true underground typology are the Atrium/courtyard concept and the Penetrational type where earth covers the entire house, except where it is retained for windows and doors for cross-ventilation opportunities and access to natural light from more than one side of the house (figure 8).

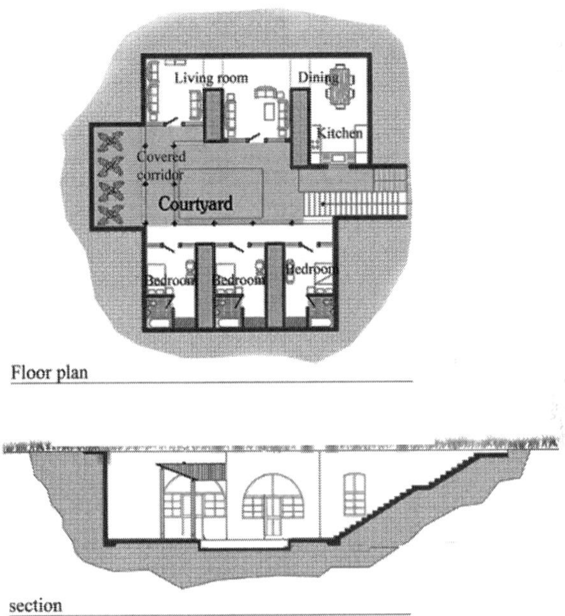

Figure 8. Underground earth shelter design

One of the most significant earth sheltered buildings in modern times is the Aloni House (figure 9). It was built in Antiparos Island in Greece and won the Greek Piranesi Award in 2009. The building epitomizes all that a modern time earth shelter represents. It combines all the design types mentioned above within a unique terrain. It also provided courtyard spaces with its landscape appearing to drift naturally into the courtyard thereby allowing for free solar penetration to the desired areas.

<table>
<tr><th>Factor</th><th colspan="4">Earth shelter building type</th></tr>
<tr><td></td><td colspan="2">Bermed</td><td colspan="2">Envelope/true underground</td></tr>
<tr><td>Passive solar potential</td><td colspan="2">Excellent</td><td colspan="2">Less effective</td></tr>
<tr><td>Thermal stability</td><td colspan="2">Less effective</td><td colspan="2">Excellent</td></tr>
<tr><td>Natural lighting potential</td><td colspan="2">Effective</td><td colspan="2">Less effective</td></tr>
<tr><td>Wind protection</td><td colspan="2">Less effective</td><td colspan="2">Excellent</td></tr>
<tr><td>Noise protection</td><td colspan="2">Less effective</td><td colspan="2">Excellent</td></tr>
<tr><td>Visual convenience</td><td colspan="2">Excellent (one directional view)</td><td colspan="2">Poor (allows only open sky view)</td></tr>
<tr><td>Appropriate Climate</td><td colspan="2">Effective for temperate</td><td colspan="2">Most effective for tropical</td></tr>
<tr><td rowspan="2">Structural cost</td><td>Modern design</td><td>Vernacular design</td><td>Modern design</td><td>Vernacular design</td></tr>
<tr><td>Intermediate</td><td>Less expensive</td><td>Most expensive</td><td>Least expensive</td></tr>
</table>

Table 1. Comparing efficiency values of the earth shelter building typology

Figure 9. Images of the Aloni House. (a) view from the hill top, (b) view from the top of the house, (c) opening leading to the courtyard, (d) the central courtyard, (e) interior view of the living room, (f) interior view from the kitchen. (Images by Julia Klimi)

4. Evaluation of energy conservation principles in earth shelter schemes

The most significant value of earth shelters and the basis for the exploitation of earth in energy saving building initiatives is its energy preservation potential. This is based on several unique physical characteristics of earth. As stated earlier, the dependability of earth in energy conservation designs is related to the natural principles of annual heat storage; huge temperature differential between the ground temperatures and the outdoor air temperatures and the insulation properties from direct solar radiation. In the cold climates, the significant property is the reduction of heat loss due to conduction through the building envelope. The amount of heat lost in this manner is a function of the thermal transmission coefficient (R-factor) of the envelope and the temperature difference between the inside of the envelope and the outside. While the R-factor for earth is substantially lower than that of other insulating materials, the large amount of earth inherent in earth sheltering can provide an overall R-factor comparable with more highly insulated structures [12].

According to investigations in [12], the temperature differential for conventional above ground structures is the difference between the outside air temperature and the interior temperature maintained for the comfort of its inhabitant. Under extreme conditions, this differential can be as much as 32°C. However, since the daily and seasonal fluctuations of temperature below the surface of the ground never equals that of the air above, therefore the deeper the temperature is taken, the less severe will be the variation. This reduced temperature differential results from the thermal storage capabilities of the soil which moderate extremes of temperature and create seasonal intervals, wherein energy from one season is transferred to the next season as in the principle of PAHS.

4.1. Solar radiation and energy conservation in earth sheltered houses

It is common knowledge that the sun is one of the most significant determinants in energy efficient building design. The radiant energy from the sun can be used as both active and passive heat generators for a building. Generally in colder climates, the active solar receptor system is oriented directly to the south, whereas all passive solar collection methods are based on trapping the radiant energy of the sun which enters through the openings on the building envelope. In the case of earth sheltered houses, the best site orientation (in cold climates) is the south-side orientation which maximizes the presence of all of the window openings whereas the remaining sides of the building are completely earth covered. The use of passive solar collection in combination with other energy conservation values is a very desirable energy efficient concept in buildings since it does not involve the capital expense that an active solar collector does. Conversely, it is important to note that, while solar radiation is desirable in the heating season of cold climates, they are not as efficient in the cooling season of hotter climates. The effect of wind on the orientation of an earth sheltered structure is a serious energy consideration [13]. Since direct exposure to cold winter winds increases heat loss due to infiltration which consequently creates a wind chill effect, it is desirable to protect a building as much as possible from this exposure. In the north hemisphere the prevailing winter winds are from the northwest. Minimizing window and door openings on the north and west sides of the house in this region will enhance energy performance.

4.2. Effects of seasonal thermal storage systems on energy conservation in earth sheltered houses

A seasonal storage system can broadly be defined as one which stores energy in one season and delivers that energy in another season. Naturally for seasonal storage systems that function as solar thermal collectors, this means that energy is collected in periods of high radiation as is the case in summer seasons and delivered in winter seasons during periods of low radiation. However to further improve the efficiency of any of the seasonal thermal storage systems, very effective above-ground insulation or super insulation of the building structure is required to minimize heat-loss from the building, thereby improving the amount of heat that needs to be stored and used for space heating.

There are three major types of seasonal (annualized) storage systems that are classified as effective or beneficial to earth shelter buildings. These are:

4.2.1. Low temperature systems:

This system utilizes the earth (soil) adjoining the building as a low-temperature seasonal heat store, thereby reaching temperatures similar to average annual air temperature while drawing upon the already stored heat for space heating. These systems can also be seen as an extension to the building design itself as the design involves some simple but significant differences when compared to conventional above ground buildings.

4.2.2. Warm temperature inter-seasonal heat system:

This also uses soil to store heat, but utilizes active solar collection mechanisms in summer to heat up thermal banks (earth mass) in advance of the heating season. Warm temperature heat stores are generated from low-temperature stores in that solar collectors are used to capture surplus heat in summer and actively raise the temperature of large mass of soil so that heat extraction is made cheaper in winter.

4.2.3. Passive annual heat storage system (PAHS):

With the development of modern passive solar building design, during the 1970s and 1980s a number of techniques were developed that enabled thermally induced and moisture-protected soil to be used as an effective seasonal storage medium for space heating, with direct conduction as the heat return method. The concept of Passive Annual Heat Storage (PAHS) is such that solar heat is directly captured by the structure's spaces and surfaces in summer and then passively transferred through its floors, walls and roof into adjoining thermally-buffered soil by conduction. It is then passively returned to the building's spaces through conduction and radiation as those spaces cool in winter. This idea was originally introduced by John Hait [11]. It includes extensive use of natural heat flow methods, and the arrangement of building materials to direct this passive energy from the earth to the building, all without using equipment. PAHS is believed to be one of the most significant ideas for energy conservation in earth sheltered buildings.

Concept of passive annual heat storage system (PAHS):

Globally, the earth receives electromagnetic radiation from the sun which is typically defined as short-wave radiation and emits it at longer wavelengths known typically as long-wave radiation. Figure 10 below shows an analysis of the earth's shortwave and long-wave energy fluxes produced with details from [14]. This absorption and re-emission of radiation at the earth's surface level which forms a part of the heat transfer in the earth's planetary domain yields the idea for the principle of PAHS. When averaged globally and annually, about 49% of the solar radiation striking the earth and its atmosphere is absorbed at the surface (meaning that the atmosphere absorbs 20% of the incoming radiation and the remaining 31% is reflected back to space). This absorbed 49% of the solar radiation presents a premise for energy efficiency in building design. The concept of earth shelter design focuses fundamentally on the utilization of the absorbed/retained heat from this annual absorption and re-emission of radiation for indoor thermal environment control.

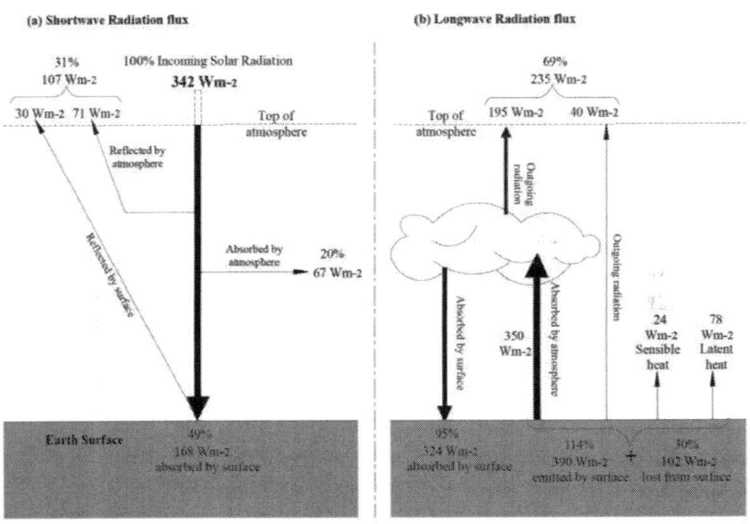

Figure 10. Earth's energy budget diagram showing the short-wave (a) and long-wave (b) energy fluxes

4.3. Analysis of soil thermal performance in earth shelter designs

The thermal property of an earth-shelter soil is an essential factor in determining its performance against other conventional above-grond houses. Due to the relatively stable temperature of the soil, the earth shelter house in summer loses heat to the cool earth rather than gaining heat from the surrounding air, and in winter the relatively warm soil offers a much better temperature environment than the subzero air temperatures. This concept is clearly confirmed by examination of the daily and yearly soil temperature fluctuations at various depths. Daily fluctuations are virtually eliminated even at a depth of 20 cm of soil. At greater depths, soil temperature responds only to seasonal changes, and the temperature change occurs after considerable delay [15]. A reasonable level of soil study is necessary in order to facilitate the comparison of the energy needed for construction (soil excavation, dewatering and concrete works) with the energy to be saved in the long run, conditions related to the insulation efficiency of the soil [16]. However the expected efficiency varies with the soil type and its water content which in some cases may have a marked effect on the thermal properties of the soil. The figure below (Figure 11) presents a typical relationship between the annual air temperatures and corresponding temperature fluctuation below the ground surface.

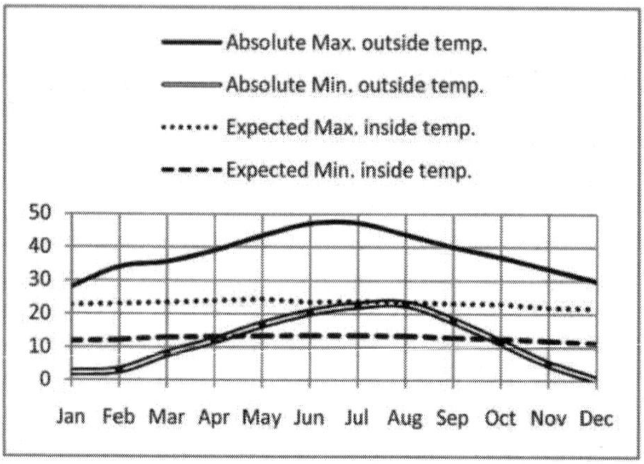

Figure 11. Annual temperature fluctuations in Riyadh from below zero to 48 ◦C and expected temperature fluctuation at 3.0m below ground level between 14◦C and 24◦C. (Data taken from [16])

In earth shelter houses, the overlaying thick earthen layer around much of the building effectively eliminates possibilities of infiltration through the building skin (as is the case in conventional above-ground houses). This can contribute significantly in reducing energy loss due to infiltration, except only through the exposed portions of the structure. Apart from the reduction of infiltration, studies identified that the application of thermal coupling of the earth-soil to the building wall places significant values to the thermal conditions of the earth shelter environment in winters. This process allows for improved thermal storage through the soil into the building walls. Since majority of modern earth shelters are built with concrete which possesses a large thermal storage capacity which can absorb the excess energy from the earth-soil, this absorbed heat is naturally released back into the building whenever the indoor air temperature is below that of the thermal mass. This thermal absorption and releasing process can provide essential heat energy required in the house for days without mechanical heating. The effect of this process is presented below (figure 12) in a thermal investigation study of a berm-type earth sheltered house in Missouri (US) covering a 4 day assessment period under a 6-hourly measurement interval [12].

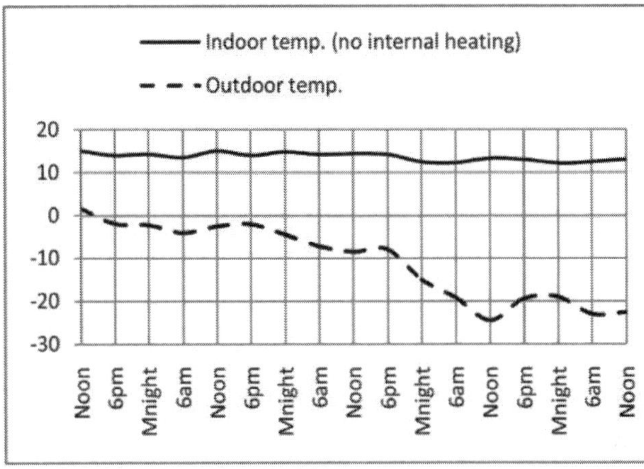

Figure 12. Temperature stability graph of an earth sheltered house in Missouri (Data taken from [12])

Determining the thermal performance of the soil for earth shelter construction involves assessing the long-term subsurface environment and above-ground temperature data. Consequently, this requires accurate environmental information on the boundary conditions, one of which is the temperature of the surrounding soil. For instance, in the case of a single basement study, a change in the mean annual ground temperature from 10◦C to 6 ◦C caused a 36% increase in heat loss [17]. Therefore, accurate data regarding diurnal and annual variation of soil temperatures at various depths is necessary to accurately predict the thermal performance of earth sheltered structures.

Study shows that actual data on soil temperatures is not usually abundant. However research has facilitated the evaluation of the underground climate in order to assess the suitability of earth sheltered structures. Algorithms for this calculation of the soil temperatures at various depths have already been developed based on existing field measurements in different regions of the world and by this, the annual pattern of soil temperatures at any depth can be accurately considered as a 'sine' wave about the annual average of the ground surface temperature. Accordingly, a mathematical method was developed to predict the long-term annual pattern of soil temperature variations as a function of depth and time for different soils and soil properties that are stable over time and depth [10]. This method is sufficiently accurate in the case certain thermal and physical characteristics are accurately estimated. The equation for estimating subsurface temperatures as a function of depth and day of the year is as follows (with the unit of cosine expressed in rad):

$$T_{(x,t)} = T_m - A_s e^{-x}\sqrt{\pi / 365\alpha}\ \cos\left\{\frac{2\pi}{365}\left[t - t_0 - \left(\frac{x}{2}\right)\left(\sqrt{\frac{365}{\pi\alpha}}\right)\right]\right\} \tag{1}$$

Where:

$T_{(x,t)}$ = subsurface temperature at depth x(m) on day t of the year (◦C),
T_m = mean annual ground temperature (equal to steady state) (◦C), as the annual temperature amplitude at the surface (x = 0) (◦C),
x = subsurface depth (m),
t = the time of the year (days) where January 1 = 1 (numbers),
t_0 = constant, corresponding to the day of minimum surface temperature (days),
α = the thermal diffusivity of the soil (m^2/day)

Through this equation, the resulting temperature profile at different depths can now be graphed and compared with the annual average air temperatures. Following the evaluation of the subsurface climate, the calculated soil temperatures can then be used to calculate the heat flux through the building surfaces. The energy efficiency of a wall in contact with the earth at varying depths can thus be investigated for local climatic conditions. This can be done by simulating the heat transfer through a subsurface wall at varying depths using a computer program, and comparing the results with an above-ground wall using the same method. This procedure is a typical preliminary assessment method with minimal input required. The expected results from the simulations provides preliminary insight into the magnitude of reduction of heat flow that the building soil climate can provide in comparison to the above-grade climate and the analysis also provides a faster approach for determining the optimum depth placement for an earth sheltered building.

Although this theory seems rightly beneficial to the energy conservation concepts in earth shelter house construction, it is also right to consider other detrimental factors like the soil's heat and cooling losses due to normal thermal transmittance factors. Earth shelters are subjected to heat and cooling losses partly via the soil to the external air, via the soil to the groundwater below or directly to the groundwater. The quantity of loss is calculable in this case and the equation is generated in [18] as follows:

$$QT = Atotal\frac{\left(v_i - vOT\right)}{RAL} + \frac{v_i - vGW}{RGW}\left[W\right] \tag{2}$$

Where:

ϑOT = mean outside temperature
≈ 0 to -5°C ≈ (ϑe + 15K)
ROT = **R**i + **R**λA + **R**λB + **R**e = equivalent resistance to thermal transmission room-outside air.
RλA = equivalent resistance of the soil to thermal conductivity.
RλB = resistance of building component to thermal conductivity.

RGW= **R**i + **R**λB + **R**λs = equivalent resistance to thermal transmission room-groundwater.
Rλs = T/λs = thermal conductivity resistance of soil to groundwater.
D = depth of groundwater
λs = thermal conductivity coefficient of soil
≈ 1.2 W/mK
ϑGW= groundwater temperature = 10°C.

4.4. Energy conservation values in earth shelter design

Earth is a great moderator of temperature change. When warmed up, it can stay warm a long time without losing much of its heat [9]. Earth does not react as fast to temperature change as air does. This means that for instance if air surface temperatures ranges from -15°C to 35°C through the year (winter through summer), then about 3 meters below, the temperature of the earth will vary only between 10°C to 15°C. This short range in difference explains the ability of earth to maintain stable temperatures throughout the year. This is a significant energy conservation tendency in the case of reducing the load on home heating and air-conditioning systems. With regards to total operating cost (excluding estimates from heat-recovery systems), energy savings of up to 60% to 70% may be realized in residential scale structures within mid-temperate zones. Instances of this were presented in [19] from the energy cost studies undertaken in [20]. In this study, a conventional 135 sq m (9m x 15m) single level residence with a hypothetical subsurface structure of the same dimension was compared. With the use of climate data and energy rates of Denver metropolis in Colorado, the study establish that the underground house provides a 72% energy savings over the surface dwelling (Table 2, 3 and 4).

Measured unit	**Conventional surface house**	**Earth shelter house**
Heat loss in winter (B.T.U. per hour)	39,927	12,720
Heat gain in summer (B.T.U. per hour)	44,650	0

Table 2. Evaluation of rates of heat loss and gain in a typical above ground house and an earth sheltered house [20]

Measured unit	**Conventional surface house**	**Earth shelter house**
Winter:		
Gas (m^3)	2,656.9 m^3 ($65.80)	871.5 m^3 ($27.60)
Oil (gal.)	710 ($129.90)	233 ($42.60)
electricity (kwh)	23,157 ($428.80)	7,596 ($191.10)
Summer		
Electricity (kwh)	3,962 ($98.40)	0

Table 3. Evaluation of annual energy consumption cost in a typical above ground house and an earth sheltered house [20]

Building type	Gas	Oil	Electricity
Above ground design (AGD)	($395)	($459)	($758)
Earth sheltered design (ESD)	($120)	($135)	($283)
Cost conservation comparison between ESD and AGD	30%	29%	37%

Table 4. Evaluation of annual cost of environmental control requirements in a typical above ground house and an earth sheltered house [20]

5. Soil suitability analysis for earth sheltered building construction

As already discussed earlier, not all soil types are efficient in use for earth sheltered building construction. The choice of construction site is mainly determined by the soil type available in a given geographical area for issues of safety against landslides and other moisture originated hazards. Some types of soil are more suitable than others in the construction of sub-grade buildings. The strength of the soil must be determined for the proposed depth of building below ground level. Though may be desirous, excavations in a very strong soil may be difficult and in the case of rocky ground, may prove impossible. On the other hand, in very weak soils the excavations are easy. In the first two cases, the capital cost and the energy expenditures involved in construction need careful examination [21]. For the third case, however, the excavation may be difficult because high lateral earth pressure requires construction of heavy walls (retaining walls), preferably made of reinforced concrete, which implies increased capital costs and energy consumption. In modern earth sheltered home construction, compaction and permeability values are the most essential standards considered in the backfill process when building a berm or elevational type construction. This is mostly due to the dangers of soil drainage. It has been noted earlier that soil-water content has distinctive effect on the thermal properties of the soil hence may affect the overall energy performance of earth-homes. Choosing a site where the water will naturally drain away from the building is the best way to avoid water pressure against underground walls. In order to improve the energy performance of the earth-soil in temperate, humid or arid tropical scenarios, drainage systems must be designed to draw water away from the structure to reduce the frequency and length of time the water remains in contact with the building's exterior. Survey has identified that ideal sites are those of hilly or mountainous terrain. The partially buried (bermed-elevational) earth-sheltered home is identified as most suitable for maximizing passive-solar heating in cold climates, however since water tends to drain down the hill toward the building and off the roof toward the back of the home, it is advisable to build in highly water-permeable soils and to install water drainage systems around the perimeter of the buried walls. Hydrology discusses infiltration as the rate at which water passes into the soil. This is also affected by the ratio of macro to micro-pores of the soil in question. The more macro-pores a soil has the easier it is for water to soak into it and drain away. Soils with coarse particles like sand or gravel or nutty or block soil structures have a high proportion of macro-pores and as a result have high infiltration rates. Soils such as clays have a high proportion of micro-pores and therefore have low infiltration rates. Figure 14 below illustrates different infiltration rates based on soil structure and texture [22].

Through the analysis below, it could be said that a good earth home design site with natural drainage also requires permeable soils. The most permeable soils as identified above are the granular type which consists of a fair amount of sand or gravel while soils with high clay content are less permeable as they expand and contract as moisture levels fluctuate. Nonetheless, it is advisable to perform percolation tests on the construction site's soil to determine the earth shelter soil permeability before construction.

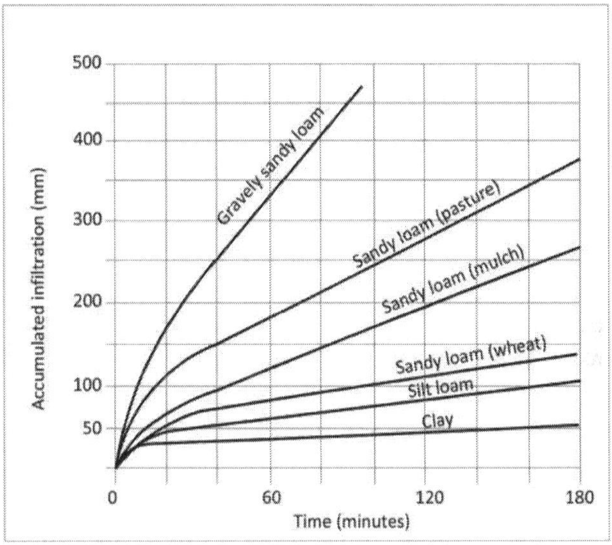

Figure 13. Infiltration curves for different soil textures

6. Thermal Integrity Analysis of earth sheltered houses

Thermal integrity factor (TIF) is a combined system for evaluating and comparing the energy performance values of building types. It is expressed in units which allow for direct comparison among such criterion as heating, ventilating, and air conditioning systems as well as the effect of various climatic conditions on different housing types. The standard unit for measuring thermal integrity values is Btu/ft^2 per degree day of the provided space condition. A TIF of 7.5 Btu/ft^2 per heating degree day is considered as representative of a baseline-factor for moderately insulated houses [23], while values in the ranges of 0.6 to 1.1 Btu/ft^2 per heating degree day are predicted for super-insulated houses [24]. Early indication of the performance of earth sheltered buildings against the conventional above-ground ones were recorded as far back as the late 1970s and 80s. Measurements were conducted on existing earth sheltered houses in some US cities. In one of the houses located in South Dakota which was monitored during 1978 and 1979, it consumed about 28,000 Btu/ft^2 for 8144 heating degree days, which yields a TIF of 3.5 Btu/ft^2 per heating degree day. The report on this house went on to note that typical above-ground framed homes in the same location generally required about 10 to 12 Btu/ft^2 per heating-degree day. This displays a

70% difference in the TIF of these two homes in the same location. Figure 13 below shows the comparative energy consumption for the above-ground and earth sheltered homes. In some other cases, earth sheltered houses display TIFs of 0 (zero) Btu/ft^2 per heating degree day. Below (table 5) are the results of the TIFs for five different buildings in Minnesota all of which recorded TIFs of less than 4.0 [25].

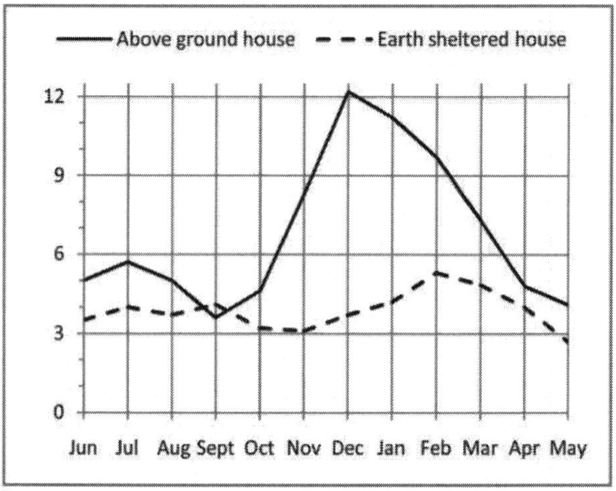

Figure 14. Comparison of monthly total energy usage in conventional above-ground and earth sheltered homes (taken from [12])

House	June 1980	July 1980	Aug. 1980	Sept. 1980	Oct. 1980	Nov. 1980	Dec. 1980	Jan. 1981	Feb. 1981
Burnsville	*nil*	*nil*	*nil*	0.65	0.84	*nil*	*nil*	*nil*	2.03
Camden	0	0	0	0	0.89	1.20	2.65	1.92	*nil*
Seward	0	0	0	0	0	2.14	3.60	2.53	3.19
Wild River	0	0	0	0	0.19	2.05	1.08	0.91	1.27
Willmar	*nil*	*nil*	*nil*	2.28	2.34	1.23	2.72	2.01	*nil*

'*nil*' = No data taken.

Table 5. Monthly thermal integrity factor for five Minnesota earth-sheltered residences

7. Conclusion

In this study, the following factors were analyzed in the hope of throwing light into the common questions that arise in the discourse of earth sheltered housing:

a. Energy conservation elements for earth shelter housing,
b. Thermal integrity values,
c. Techniques for maximizing the thermal loads necessary for comfort conditions in passively heated or passively cooled earth shelters,

d. Soil suitability, depth of placement and design techniques that optimizes structural integrity in earth sheltered house construction.

This study also presented some of the valuable analysis and results in earth shelter building evaluations as premise for assessing the potentials of passively heated earth sheltered houses. This is achieved through a review of previous performance assessments of monitored conditions in existing earth sheltered buildings. Through this review, thermal integrity factors (TIF) of existing earth sheltered homes were identified, which when compared with other housing types, perform significantly better than conventional above-ground dwellings. It also looked at both summer and winter impacts of earth shelter house types utilizing the passive approach under the different climate conditions. This study identified that the thermal integrity value of passively heated earth sheltered house is comparable with other energy-efficient approaches such as super-insulated and passive solar constructions which are much better in energy conservation performance than the conventional above-ground constructions. It further presents the criteria for identifying the appropriate soil type (sub-grade materials) needed in building earth sheltered houses with passive thermal approach. These are categorized under thermal inertia properties, bearing capacity and drainage properties. Based on the available information to date, it can be said that earth sheltered houses maintain heating energy consumption that is lesser by up to 75%. This claim appears to be substantiated as earth sheltered house compared to conventional above-ground house presents a lesser calculated or monitored TIF.

Having looked through the benefits and potentials of earth and the overall understanding of its potential for energy conservation through earth-sheltered construction, it is hoped that this review contributes to the information available so far on means of assessing the performance of earth shelters and associated thermal properties that affects it. It is then possible for designers and planners in different regions to have access to a simple framework for assessing its efficiency at the initial planning stages. The resulting outputs can then be used for the heat transfer and energy conservation analysis within the building units. Results from this analysis will provides insight into the degree of passive heating and cooling or reduction in heat flow that the soil climate can provide as compared to the surface climate as well as suggesting parameters for depth placement of earth shelter buildings for more efficient results.

Author details

Akubue Jideofor Anselm
Architecture Department, University of Nigeria, Nigeria

8. References

[1] Carpenter P, (1994) Sod It: An Introduction to Earth Sheltered Development in England and Wales, Coventry University, Coventry,.

[2] Carmody J, Sterling R, (1984) Design considerations for underground buildings, Underground Space 8: 352-362.

[3] Kumar R, Sachdevab S, Kaushik S.C, (2007) Dynamic earth-contact building: A sustainable low-energy technology, Building and Environment 42: 2450-2460.

[4] Dodd J, (1993) Earth sheltered settlements, a sustainable alternative, in: Proceedings of the Earth Shelter Conference, Coventry University 3: 26-36.

[5] Moreland F.L, (1975) An alternative to suburbia, in: Proceedings of the Conference on Alternatives in Energy Conservation: The Use of Earth-covered Buildings, National Science Foundation, Fort Worth, TX,.

[6] Golany G. S, (1983) Earth Sheltered Habitat (History, Architecture and Urban Design), Van Nostrand Reinhold Company Inc., New York.

[7] Rahman M.M, Rasul M.G, Khan M.M.K, (2010) Energy conservation measures in an institutional building in sub-tropical climate in Australia. Applied Energy 87: 2994-3004

[8] Reynolds M, (1991) Earth-ship Systems and Components, Solar Survival Press

[9] Wells M. B, (1975) To Build without Destroying the Earth. Alternatives in Energy Conservation: The Use of Earth Covered Buildings. Washington, D.C. U.S. Government Printing office. Pp. 211-232

[10] Labs K, (1979) Underground building climate, Solar Age 4: 10 44-50.

[11] Hait J, (1983) Passive Annual Heat Storage: Improving the Design of Earth Shelters, Rocky Mountain Research Center

[12] Wendt R. L, (1982) Earth-Sheltered Housing: An Evaluation of Energy-Conservation Potential. U.S. Department Of Energy, Oak Ridge Operations, TN. pp. 8-18,

[13] Minnesota University (1979). The Underground Space Center, Earth Sheltered Housing Design. Minneapolis, Minnesota: Van Nostrand Reinhold Company. pp. 20

[14] Bonan G, (2002) Ecological Climatology: Concepts and Applications, Cambridge Press, United Kingdom

[15] US Department of Housing and Urban Development, (1980). "Earth Sheltered Housing" Code, Zoning, and Financing Issues

[16] Khair-El-Din A. M, (1991) Earth Sheltered Housing: An Approach to Energy Conservation in Hot Arid Areas Architecture and Planning Riyadh. 3:3-18

[17] Mitalas G.P, (1982) Basement Heat Loss Studies, DBR/NRC, Ottawa.

[18] Klaus D, (2003) Advanced Building Systems: A Technical Guide for Architects and Engineers, Published for Architecture Basel, Boston, Berlin. pp. 50.

[19] Labs K, (1975) The Architectural Use of Underground Space: Issues and Application. Master's Thesis/Washington University, May, Mechanicsville PA. pp. 121

[20] Harrison L, (1975) Is it time to go Underground? The Navy Civil Engineer, Fall. pp. 28-29

[21] Henna A. M, (1980) Building Underground Alternatives. Miami published research for the Energy Conservation Conference, Florida.

[22] McLaren R. G, Cameron K. C, (1990) Soil science. An introduction to Properties and Management of New Zealand Soils. Oxford university press, Auckland.

[23] Lewis D, Fuller W, (1979). Solar Age, p. 31

[24] Shurcliff W. A, (1980). Super-insulated Houses and Double-Envelope Houses, A Preliminary Survey of Principles and Practice, 2nd Ed., Cambridge, Mass. p. 6.05.

[25] Goldberg L. F, Lane C. A, (1981). A Preliminary Experimental Energy Performance Assessment of Five Houses in the MHFA Earth-sheltered Housing Demonstration Program, University of Minnesota, Minneapolis. pp. 10.

Chapter 6

Low Energy-Consumption Industrial Production of Ultra-Fine Spherical Cobalt Powders

Chong-Hu Wu

Additional information is available at the end of the chapter

http://dx.doi.org/10.5772/52113

1. Introduction

Cobalt powders have been used extensively in cemented carbides, high-temperature alloys, PCD and PCBN, and magnetic materials, etc., due to its excellent physical, chemical, and mechanical properties. The fabrication and final properties of the above materials are strongly affected by the quality (purity, phase, size, shape, dispersity, fluidity, etc.) of Co powders. In order to fabricate a homogeneous and densified microstructure without pores, Co-pool, Co-poor and Co-free zones, etc., ultra-fine spherical Co powders have been desired with the development of ultra-fine and even nano grain materials. However, it is very difficult to industrial fabricate ultra-fine spherical Co powders with good quality by the conventional decomposition and hydrogen-reduction technology. In this paper, a new low energy-consumption industrial production technology-a continuously dynamic-controlled combustion synthesis (CDCCS) technology has been proposed about investigating how to obtain ultra-fine spherical Co powders (the average particle size is smaller than 0.8μm, and the length-diameter ratio is smaller than 2) with a lower impurity content.

2. Cobalt: Properties, minerals, extraction and applications

2.1. Properties

2.1.1. Physical properties

Cobalt does not occur naturally as a pure metal, but is a component of more than near a hundred naturally occurring minerals, including various sulfides, arsenides, sulfoarsenides, hydrates, and oxides. Pure cobalt can be produced by reductive smelting, and was firstly prepared by G. Brandt in 1735. Cobalt was confirmed as an element by T. Bergman in 1780. Cobalt is a metallic transition element, and its position in the Periodic Table is characterized:

Name, symbol, number: Cobalt, Co, 27

Element category: Transition metal

Group, period, block: 9(VIIIA), 4, d

Standard atomic weight: 58.933195

Electron configuration: 4s3d7

Electrons per shell: 2, 8, 15, 2

Co is a brittle, hard metal, resembling iron and nickel in appearance. Pure cobalt produced by reductive smelting is a hard, lustrous, silver-gray metal. Co has a relative permeability two thirds that of iron [1]. Its Curie temperature and magnetic moment are 1115°C [2] and 1.6~1.7 Bohr magnetons/atom [3], respectively. The basic physical properties are listed in Tab.1, and vapor pressure vs. temperature is shown in Fig.1. The transformation is sluggish and accounts in part for the wide variation in reported data on its physical properties. Metallic Co occurs as two crystallographic structures: a hexagonal closed-packed crystal structure (hcp) and a face-centered cubic crystal structure (fcc). During cooling, Co usually undergoes a polymorphous transformation from fcc to hcp. The temperature corresponding to the equilibrium between the high-temperature β (fcc) and low-temperature α (hcp) phases in Co is 417°C [4], but in fact, the energy difference is so small that random intergrowth of the two is common [5].

Physical properties	values
Color	Silver-gray
Density	$8.9 g/cm^{-3}$
Liquid density at melting point temperature	$7.75 g/cm^{-3}$
Melting point	1768K, 1495°C, 2723°F
Boiling point	3200K, 2927°C, 5301°F
Heat of fusion	16.06kJ/mol
Heat of vaporization	377kJ/mol
Molar heat capacity	24.81J/(mol·K)

Table 1. The basic physical properties of Co metal

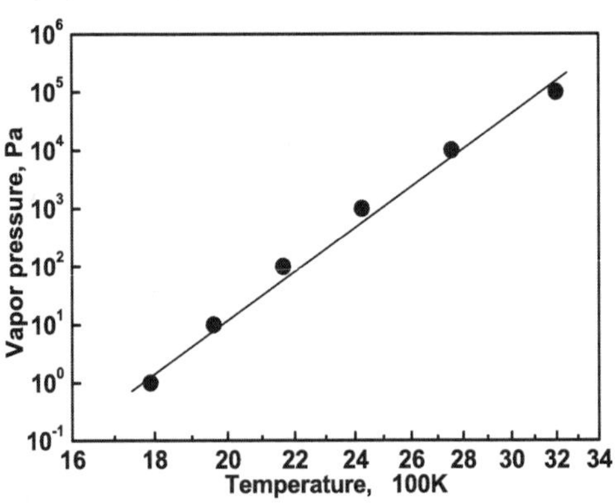

Figure 1. Vapor pressure of Co

2.1.2. Chemical properties

Pure Co does not dissolve in water and soil, and can stay for years at room temperature (RT). Co can stay in the air for a few days, and can be oxidized to CoO at about 300°C, and can be burnt to Co_3O_4 at high temperature. Fine Co powders prepared by the hydrogen-reduction technology are easily oxidized to cobalt oxide and even generate self-ignite in the air. So the fine Co powders must be carefully preserved in a vacuum or an inert gas container.

Co is a weakly reducing metal, and is easily attacked by halogens and sulfur, and is also rapidly dissolved by HCl, H_2SO_4 and HNO_3 acid solution, and is slowly eroded by HF, NH_4OH, NaOH solution.

Co can solid-dissolve many metal and nonmetal atoms to form many intermetallic compounds. Co can well infiltrate many ceramics (WC, TaC, TiC, ZrC, TiN, Al_2O_3, cBN, diamond, etc.), and the almost all wetting angles are lower than 50° [6], so it is often used as a binder in the above ceramic-metal composites.

2.1.3. Mechanical properties

Metallic Co occurs as two crystallographic structures: hcp-α and fcc-β. Strength and hardness of hcp-α should be higher than those of fcc-β, but inverse for their plasticity because there are more slipping systems in fcc-β than hcp-α. Some mechanical properties of Co are listed in Tab.2.

Mechanical properties	Crystal structures	
	hcp	fcc
Electrical resistivity (Ω·m)	6.24×10^{-8} at 20°C	-
Thermal conductivity (W/(m·K))	100	-
Thermal expansion (/K)	13.36×10^{-6} at 25°C	-
Young's modulus (GPa)	209	<hcp-Co
Shear modulus (GPa)	75	<hcp-Co
Bulk modulus (GPa)	180	<hcp-Co
Posisson ratio	0.31	-
Vickers hardness (MPa)	1043	<hcp-Co

Table 2. Some mechanical properties of Co

2.2. Co minerals and extraction

Co is not a typically rare metal since it ranks 33 in abundance. Content of Co in the earth's crust is about 0.035wt.%, and about 2.3 billion ton in the sea. Nearly, all Co is always found associated with metallic-lustered ores of other metals (for example, Cu, Ni, Fe, Pb, Zn, etc.), and Co minerals without other metals is very less except cobaltite in Morocco. So generally it is produced as a by-product of other metals mining. Near a hundred cobalt minerals have been already found in the nature, but about only 20 cobalt minerals are valuable and available, listed in Tab.3.

Based on the complexity of the Co minerals, the extraction processing is very complicated and efficiency of recovery is also very low. As usual, Co in the minerals is firstly concentrated or is transformed into the soluble states by the pyro-refining, and then Co in the calcine of pyro-refining is further enriched and extracted by the hydrometallurgy, the finally the cobalt compounds or pure cobalt are obtained. An extraction processing is shown in Fig.2 [7]. Section (the production of Co powders) in the dotted line scope in Fig.2 will be emphasized in the chapter.

Cobalt minerals	Chemical formula	Theory cobalt contents (wt.%)	Actual cobalt contents (wt.%)	Distribution
	$CoAs_2$	23.2	15~21	Canada, Morocco, USA
	$(Co,Fe)As_2$	28.2	9~23	Morocco, Canada, Russian
Arsenides	$CoAs_3$			
	$(Co,Ni)As_3$	20.8	16~20	Morocco, Canada, Russian
	$(Co,Ni,Fe)As_3$			
	CoAsS	35.5	29~35.3	China, Canada, Morocco, USA, Australia
	(Co,Fe)AsS		15~20	Morocco
Sulfides	$CuCo_2S_4$	38.7	27~42	Zaire, Zambia
	Co_3S_4	48.7	36~53	Zaire, Zambia
	$(Co,Ni)_3S_4$	26	4~10	USA, China
	CoS_2			Zaire, China
	$m(Co,Ni)O \cdot MnO_2 \cdot nH_2O$	<32	<30	Zaire, China, New Caledonia
Oxides	$3CoO \cdot As_2O_3 \cdot 8H_2O$	29.5		Morocco, Canada
	$CuO \cdot 2Co_2O_3 \cdot 6H_2O$	57	45~47	China, Zaire
	$CoCO_3$	49.6		Zaire, Zambia

Table 3. Cobalt minerals

2.3. Applications

Due to the excellent physical, chemical, and mechanical properties, it is widely used in fabricating various alloys (which are used as high-temperature and wear resistant components, dies, saws, cutting tools, etc.) by powder-metallurgy technology, such as super alloys, high speed steels, cemented carbides, PCD (polycrystalline diamond) and PCBN (polycrystalline cubic boron nitride), etc. Co in the above alloys is usually used as an additive, an alloying element or a binder.

An important use for Co is in the field of high-temperature alloys. Required in gas turbines, jet engines, and similar applications, such alloys retain their strength above 650°C; these alloys contain 5~65wt.% Co. Even higher operating temperatures in turbines have resulted in an increased use of cobalt-containing and cobalt-based alloys known generally as super alloys. These can withstand severe operating conditions and temperatures up to 1150°C. For example, Nimonic 90 is a nickel-based alloy containing 18wt.% Co, a similar amount of Cr, and some Ti, and Waspaloy is another alloy of this type.

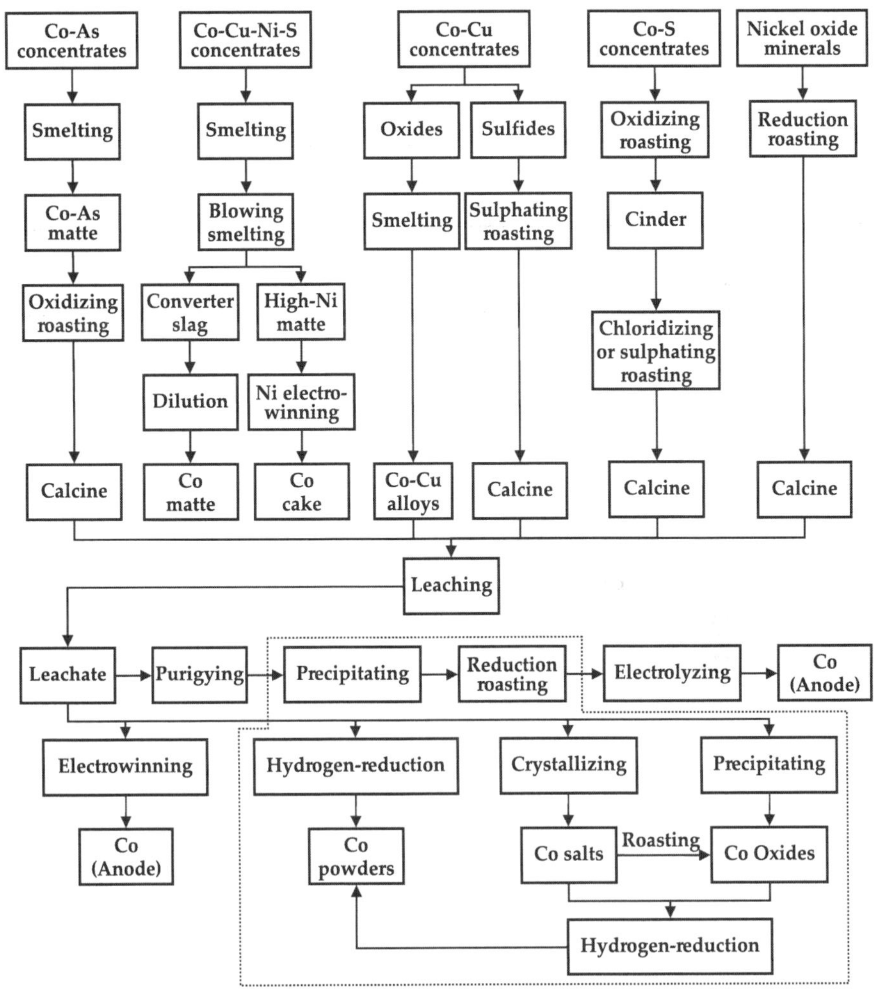

Figure 2. Cobalt extraction processing (the section in the dotted line scope will be emphasized in the chapter)

Cemented carbides, PCD and PCBN are used as cutting tools, wear-resistant components, dies, saws. In the production of a so-called cemented carbide, i.e. tungsten carbide composites, a briquetted mixture of tungsten carbide and soft cobalt powders is compacted and sintered at a temperature above the melting point of cobalt. The latter melts and binds the hard carbides, giving them the toughness and shock resistance needed to make carbides of practical value for cutting tools, drill bits, dies, and saws, etc. Co is the most satisfactory matrix metal for this purpose and may be adjusted in amounts from 3 to 25 percent by weight. A briquetted mixture of diamond or cBN and Co powders are sintered at high temperature and high pressure, and PCD and PCBN materials are obtained.

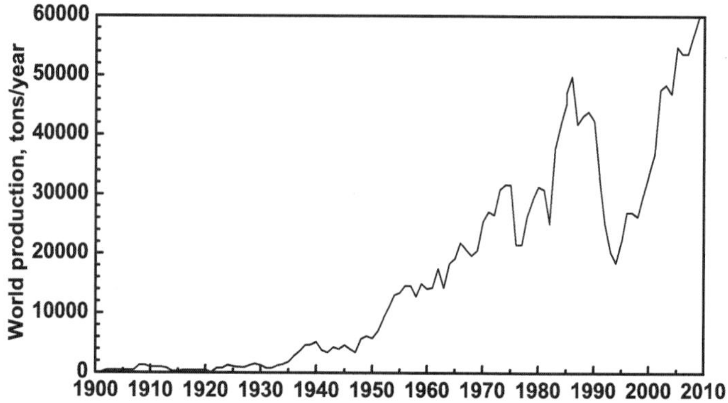

Figure 3. World production trend of cobalt

Very strong magnets are created when Co is alloyed with other metals. So Co is used in fabricating many advanced magnetic materials.

Cobalt's use in rechargeable batteries is the fastest growing use. Notably in 2007, the percentage of cobalt use for rechargeable batteries rose to 25% of total cobalt demand from 22% in 2006.

World production trend of cobalt is shown in Fig.3 [8,9]. The production of cobalt has been increasing steadily since 1996.

3. Expectation qualities of cobalt powders in powder metallurgy industry

Powder metallurgy materials (PMM) have been usually fabricated by mixing, pressing, and sintering processing. The quality (purity, phase, size, shape, dispersity, fluidity, etc.) of raw powders acts an important role in determining the fabrication and final properties of PMM. Especially with the rapid development of ultra-fine grain cutting tool materials (such as, WC/Co alloys, PCBN, PCD, etc.), ultra-fine spherical Co powders (the size is $\leq 0.8\mu m$ and the length-diameter ratio is smaller than 2) have been desired in order to fabricate a homogeneous and densified microstructure without pores, Co-pool, Co-poor and Co-free zones, etc.

Though hcp-α Co is a room temperature stable phase, fcc-β Co can also be steadily retained at RT by some especial techniques (for example, rapidly cooling). For metal alloys, the strength and plasticity are generally contrary, so hcp-α and fcc-β structures must be alternative in applications. For cemented carbides (WC/Co alloys), hcp-α Co powders are desired because the cold welding among Co particles due to its higher brittleness can be decreased during ball-milling. The cold welding (seen as "hard" agglomerations, as shown in Fig.4) among Co particles can easily result in "Co-pool" which result in a rapidly decrease of mechanical properties of WC/Co alloys. However, fcc-β cobalt in the sintered WC/Co alloys is desired by rapidly cooling during sintering because its good plasticity can improve the toughness of WC/Co alloys.

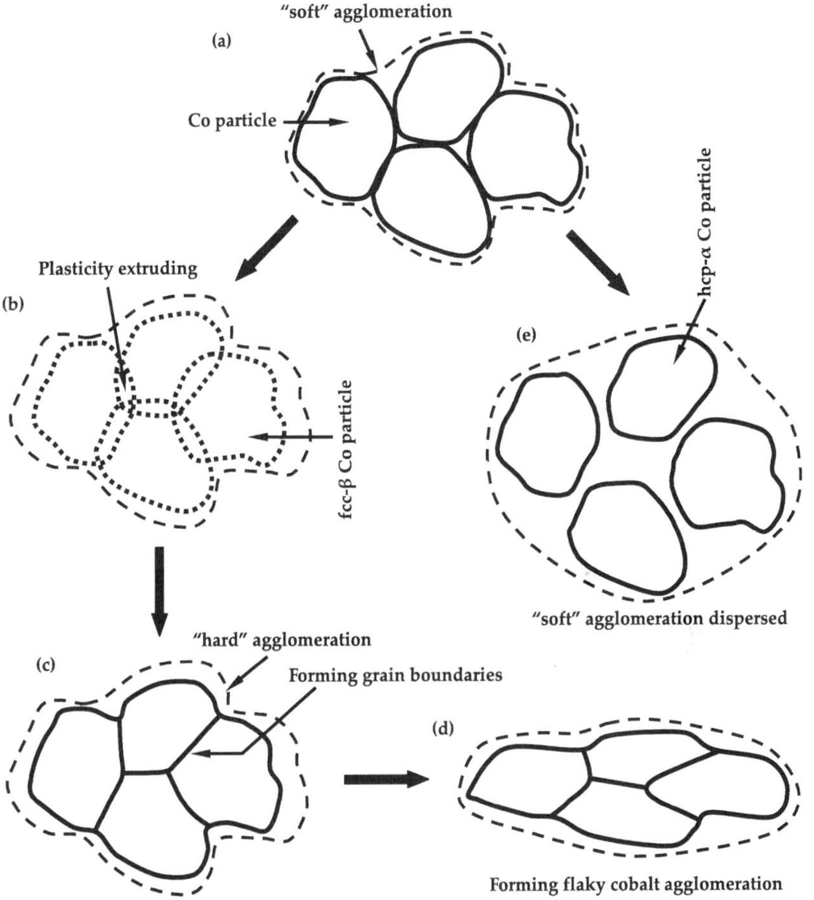

(a) Initial morphology ("soft" agglomeration) of Co particles; (b) Occurring cold welding among fcc-β Co particles due to each other extruding; (c) Forming grain boundaries at the cold welding positions, and forming a "hard" agglomeration; (d) Forming flaky Co agglomeration due to plastic deformation; (e) hcp-α Co particles well dispersed

Figure 4. Schematic representation of the evolution of hcp-α and fcc-β Co particle agglomerations during ball-milling

4. Decomposition-hydrogen-reduction to fabricate cobalt powders

Co powders in powder metallurgy industry are mainly fabricated by a decomposition-hydrogen-reduction technique using cobalt compounds of cobalt oxalate ($CoC_2O_4{\cdot}2H_2O$), cobalt carbonate ($CoCO_3{\cdot}xH_2O$), or cobalt oxide (Co_3O_4) powders as raw materials. Schematic representation of the device is shown in Fig.5. Raw powders are uniformly tiled in the boats, and then the boats are pushed into hydrogen-reduction furnace at regular intervals by a Feeder. The raw powders can also be continuously fed by a conveyor belt in some companies [10], but we consider that the conveyor belt should carry off a lot of heat, and resulting in an increase of energy-consumption and the complexity of the device is also increased. H_2 from an inverse direction enters the furnace. The raw powders are heated, decomposed, and reduced in the heating zone (in which there are usually not less than three different temperature zones). Obtained Co powders are cooled to RT in the long cooling zone, and collected into a Receiver, and protected by N_2. In order to reduce energy-consumption and protect the environment, the residual H_2 including some other gases, such as CO, CO_2, H_2O, etc., is purified by a Purifier, and then reused.

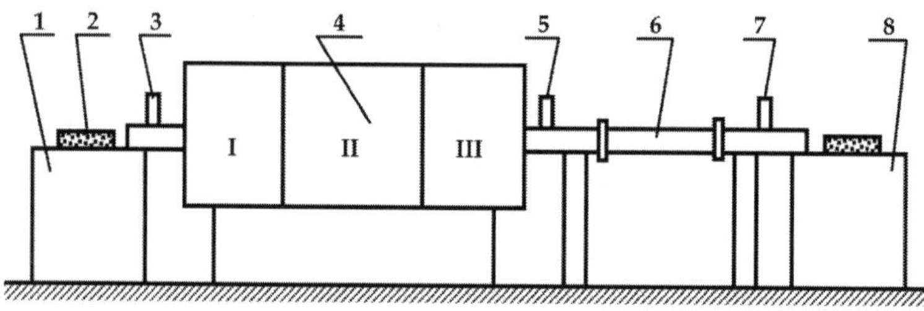

1. Feeder; 2. Boat; 3. H_2 exit to a Purifier; 4. Heating zone; 5. H_2 entrance; 6. Cooling zone; 7. N_2 Protecting; 8. Receiver

Figure 5. Schematic representation of the device used in fabricating Co powder

In order to well understand the decomposition mechanisms of $CoC_2O_4{\cdot}2H_2O$, $CoCO_3{\cdot}xH_2O$, Co_3O_4 powders and further guide the production of Co powders, their thermal decomposition kinetics in an inert gas and air will be detailedly studied and discussed in the following sections. Effect of morphology, size and purity of three raw powders and technical parameters on the properties of Co powders is also discussed. Some beneficial methods are also summarized and proposed to improve the quality of Co powders.

4.1. Cobalt oxalate

The chemical formula of the commercial cobalt oxalate is $CoC_2O_4{\cdot}2H_2O$. Fig.6 shows TG and DTA curves of $CoC_2O_4{\cdot}2H_2O$ in an inert gas (such as N_2, Ar) and air, respectively. DTA curve in the inert gas exhibits two endothermic peaks, which are accompanied by the weight loss of ~19.67% and ~48.09% in the TG curve, respectively. The ranges of reactive temperatures at the peaks are in about 170~225°C and 350~440°C, respectively, which will change a few with different testing conditions (such as changing heating rate). Such weight loss agrees with the value calculated for the following two transformations:

$$CoC_2O_4 \cdot 2H_2O \rightarrow CoC_2O_4 + 2H_2O \uparrow \; (\text{According to the first endothermic peak}) \tag{1}$$

$$CoC_2O_4 \rightarrow Co + 2CO_2 \uparrow \; (\text{According to the second endothermic peak}) \tag{2}$$

DTA curve in the air exhibits an endothermic peak and an exothermal peak, which are accompanied by the weight loss of ~19.67% and ~36.43% in the TG curve, respectively. The ranges of reactive temperatures at the peaks are in about 170~225°C and 270~320°C, respectively, which will also change a few with different testing conditions (such as changing heating rate). Such weight loss agrees with the value calculated for the following two transformations:

$$CoC_2O_4 \cdot 2H_2O \rightarrow CoC_2O_4 + 2H_2O \uparrow \; (\text{According to the first endothermic peak}) \tag{3}$$

$$3CoC_2O_4 + 2O_2 \rightarrow Co_3O_4 + 6CO_2 \uparrow \; (\text{According to the second exothermic peak}) \tag{4}$$

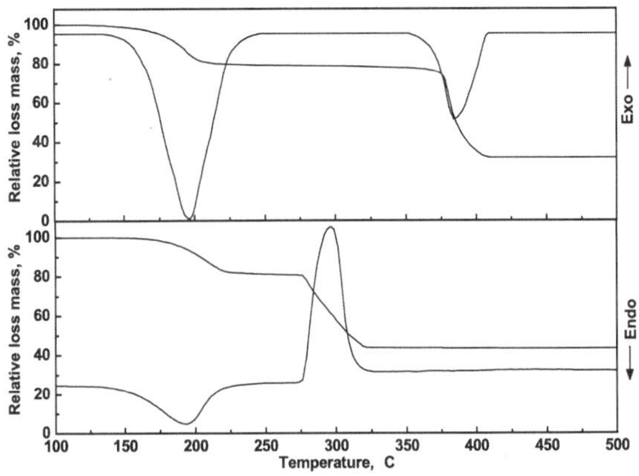

Figure 6. TG and DTA curves of $CoC_2O_4 \cdot H_2O$ in an inert gas (up) and air (down)

In actual industrial production of Co powders, the decomposition-hydrogen-reduction processing of $CoC_2O_4 \cdot 2H_2O$ occurs in a H_2 atmosphere in the device of Fig.5. Tikkanen *et al* [11] indicated that the decomposition-hydrogen-reduction processing of $CoC_2O_4 \cdot 2H_2O$ was divided into a guide stage and a decomposition stage. The dehydration (see the reaction (1-1)) dominates during the guide stage. Huang *et al.* [12] indicated that high crystallinity CoC_2O_4 crystals with less defects can be obtained under a slow dehydration rate, and then further decomposition processing changes difficulty. Therefore, we easily surmise that it is disadvantageous to fabricate ultra-fine Co powders due to a decrease of the nucleation rate. In reverse, CoC_2O_4 crystals with many defects can be obtained under a rapid dehydration rate, and then the decomposition changes easy; resulting in an increase of the nucleation rate. So we propose to increase properly the feeding rate of boat and temperature during the guide stage. In actual industrial production, we can shorten properly the scope of the "I" heating zone in Fig.5 and increase the dehydration temperature (usually higher 30~50°C than the result of DTA). On the other hand, the dehydration processing of $CoC_2O_4{\cdot}2H_2O$ can be incompletable because hcp-α Co powders are easily obtained during the decomposition of the CoC_2O_4 with a few H_2O.

Fig.7 shows the effect of decomposition temperature and time of $CoC_2O_4{\cdot}2H_2O$ on the specific surface area of Co powders [11]. There are two activation energies for the growth of Co powders, which implies two different growth mechanisms: the aggregating growth among Co crystalline nuclei at lower temperature and the second aggregating growth among Co particles. It is obvious that the second growth must be inhibited during decomposition. The temperature occurring the second growth is about ~450°C, which is similar to the results of DTA. Therefore, the temperature in the decomposition stage can not usually be higher than 450°C in order to obtain ultra-fine Co powders. In fact, it is very difficulty to completely inhibit the second growth due to the local overheating in the boat. The following methods are usually used to improve the temperature homogeneity in the boat: increasing the surface of the boat, decreasing the layer thickness of $CoC_2O_4{\cdot}2H_2O$ powders, and increasing the flow rate of H_2.

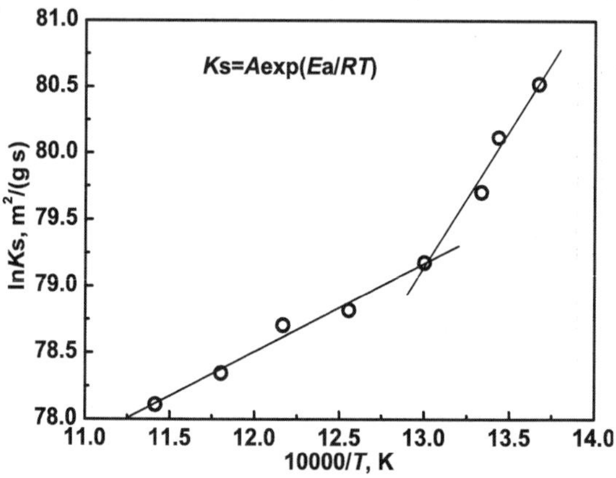

Figure 7. Effect of decomposition temperature and time of $CoC_2O_4{\cdot}2H_2O$ on the specific surface area of cobalt powders

Morphology of the commercial $CoC_2O_4{\cdot}2H_2O$ powders is a short-fibrous structure with an average diameter of about 0.5~1μm and an average length of about 4~10μm, or a bunch (3~4μm) of fibrous particles is formed when several short-fibrous cobalt oxalate particles are adhered or bonded together during crystallization, as shown in Fig.8(a). Because morphology of Co powders can easily inherit that of cobalt oxalate, it is very difficulty to obtain spherical Co powders by decomposing cobalt oxalate; usually replacing a short-string or dendritic structure, shown in Fig.8(b). Furthermore, the production practice shows that there is always 2~4% "hard" agglomeration (a size of 10~30μm) in Co powders. The decomposition processing of cobalt oxalate is schematically shown in Fig.9. The short-fibrous $CoC_2O_4{\cdot}2H_2O$ particles stacked disorderly in the boat are decomposed and reduced *in situ* in the device of Fig.5. The Co nuclei are nucleating and growing to form a short-string structure along the short-fibrous of $CoC_2O_4{\cdot}2H_2O$ particle, and the short-string or dendritic structure, and even "hard" agglomerations are formed when several short-string Co particles grow or bond together.

Figure 8. Morphology of the commercial cobalt oxalate powders and Co powders fabricated by decomposing cobalt oxalate: (a) short-fibrous cobalt oxatate powders and (b) short-string or dendritic cobalt powders

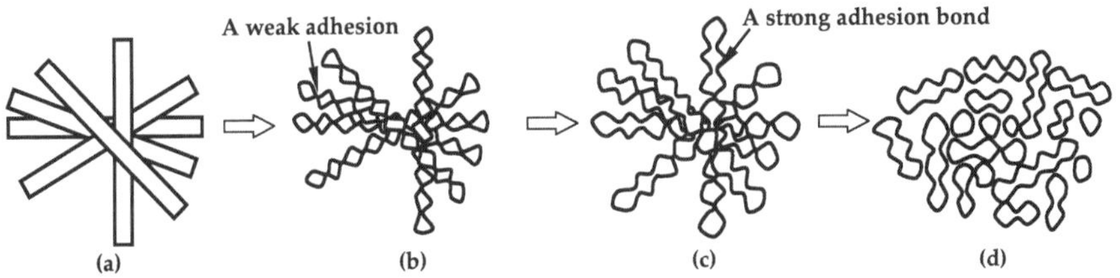

(a) Morphology of cobalt oxalate powders; (b) A short-string structure of Co particles along the short-fibrous after hydrogen-reduction, there is a weak adhesion between Co particles; (c) Co particles grow up, and a strong adhesion bond between Co particles is formed; (d) The short-string or dendritic structure of cobalt powders after sifting

Figure 9. Schematic representation of *in situ* decomposition-hydrogen-reduction processing of cobalt oxalate

In order to obtain ultra-fine spherical cobalt powders, there are two methods: to obtain spherical cobalt oxalate powders and to impede or break the adhesion of cobalt nuclei during the decomposing processing of cobalt oxalate. Spherical cobalt oxalate powders can be crystallized by adding a few spheroidizer [13,14] into the cobalt salt leachate, but the spheroidizer can deteriorate the properties of the final sintered alloys. Du *et al.* [15] proposed that spherical cobalt oxalate powders can be crystallized by adding a pulsed magnetic field in the cobalt salt leachate, but the size of cobalt oxalate particles is up to 3.5μm which is disadvantageous to fabricate ultra-fine Co powders. Huang *et al.* [16] proposed to decrease the size of cobalt oxalate powders by an airflow dispersion method, but it is obvious to largely increase the production cost and to easily result in an increase of the content of impurity in cobalt oxalate powders. Li *et al.* [17] proposed to impede the adhesion of cobalt nuclei by doping tungsten powders into cobalt oxalate powders, and then equiaxed or spherical-like Co powders are obtained and the size of Co particles decreases with increasing tungsten powders. But tungsten can promote to transform hcp-Co into fcc-Co, and the difficulty controlling the content of carbon in the final sintered WC/Co alloys is increased because the additional tungsten in Co powders must be completely reacted to form WC by adding carbon powders.

4.2. Cobalt carbonate

The chemical formula of the commercial cobalt carbonate is $CoCO_3{\cdot}xH_2O$ (x is lower than 1.). Fig.10 shows TG and DTA curves of $CoCO_3{\cdot}xH_2O$ in an inert gas (such as N_2, Ar) and air, respectively. DTA curve in the inert gas exhibits two endothermic peaks. The ranges of reactive temperatures at the peaks are in about 130~220°C and 320~420°C, respectively. According to XRD results in Fig.11, there are the following two reactions:

$$CoCO_3{\cdot}xH_2O \rightarrow CoCO_3 + xH_2O \uparrow \ \left(\text{According to the first endothermic peak}\right) \tag{5}$$

$$CoCO_3 \rightarrow CoO + CO_2 \uparrow \quad \left(\text{According to the second endothermic peak}\right) \tag{6}$$

DTA curve in the air also exhibits two obvious endothermic peaks. The ranges of reactive temperatures at the peaks are in about 130~220°C and 250~320°C, respectively. According to XRD results in Fig.11, there are the following two reactions:

$$CoCO_3{\cdot}xH_2O \rightarrow CoCO_3 + xH_2O \uparrow \ \left(\text{According to the first endothermic peak}\right) \tag{7}$$

$$6CoCO_3 + O_2 \rightarrow 2Co_3O_4 + 6CO_2 \uparrow \ \left(\text{According to the second endothermic peak}\right) \tag{8}$$

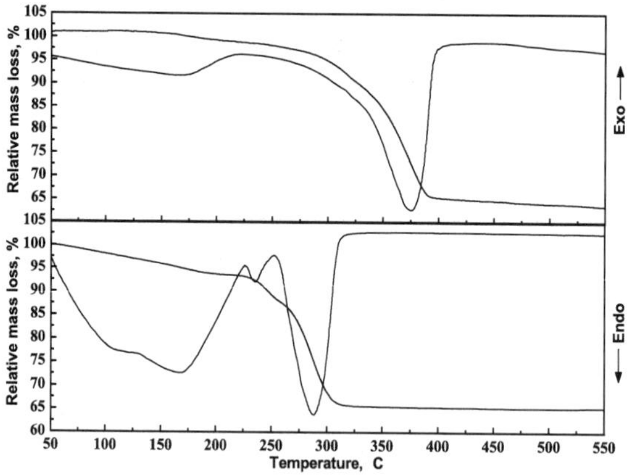

Figure 10. TG and DTA curves of $CoCO_3{\cdot}xH_2O$ in an inert gas (up) and air (down)

In addition, there seems a weak exothermic peak in about 220~270°C, in which the weight loss on the TG curve is a few slower than that in 270~320°C. According to XRD result at 220°C in the air, there also are the following two possible reactions:

$$CoCO_3 \rightarrow CoO + CO_2 \uparrow \quad (9)$$

$$6CoO + O_2 \rightarrow 2Co_3O_4 \ \left(\text{According to the weak exothermic peak}\right) \quad (10)$$

A similar result was reported in Ref. [18], namely the decomposed products of $Co_5(OH)_6(CO_3)_2{\cdot}xH_2O$ powders at a lower temperature are a mixture CoO and Co_3O_4 powders, and CoO is oxidized to form Co_3O_4 at a higher temperature.

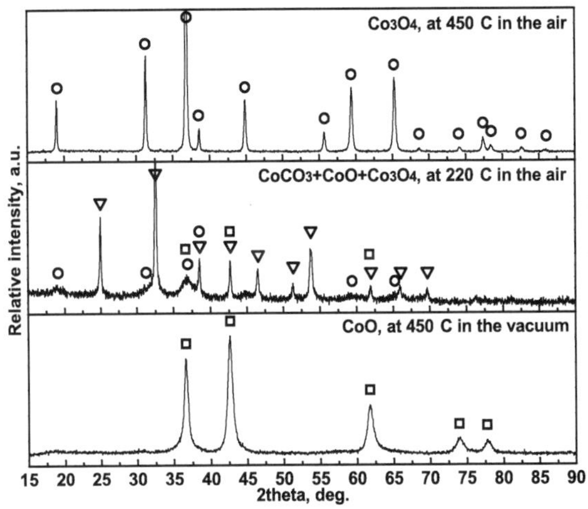

Figure 11. XRD patterns of the decomposed products of $CoCO_3{\cdot}xH_2O$ powders at various conditions

Being different from the short-fibrous $CoC_2O_4{\cdot}2H_2O$ powders, the production practice [19,20] indicates that the spherical Co powders can be easily obtained by decomposing a nearly spherical $CoCO_3{\cdot}xH_2O$ powders (see Fig.12(a)). However, the minimum average size of Co powders obtained by decomposing $CoCO_3{\cdot}xH_2O$ powders is only up to 0.8~0.9μm (see Fig.12(b)), and Co powders also possess a wide particle size distribution. The content of fcc-Co in the powders is as high as 60%.

Another disadvantage is: the content of some impurities (such as S and Ca) in $CoCO_3{\cdot}xH_2O$ powders are usually higher than those in $CoC_2O_4{\cdot}2H_2O$ powders due to the different precipitation extraction processing. It is very difficulty to completely eliminate the impurities, which can be often retained into the final sintered WC/Co alloys. And then the properties of WC/Co alloys can be deteriorated by the impurities. Fig.13 shows that there is a higher content of S and Ca at a fracture origin in a sintered WC/Co alloy, and the raw Co powders are fabricated by decomposing and reducing cobalt carbonate powders. Although the impurities can be transformed into some non-deleterious compounds by adding a few rare earth elements [21,22], the other questions are again introduced during fabricating WC/Co alloys.

Figure 12. Morphology of (a) the commercial $CoCO_3{\cdot}xH_2O$ powders and (b) Co powders fabricated by decomposing and reducing $CoCO_3{\cdot}xH_2O$ powders

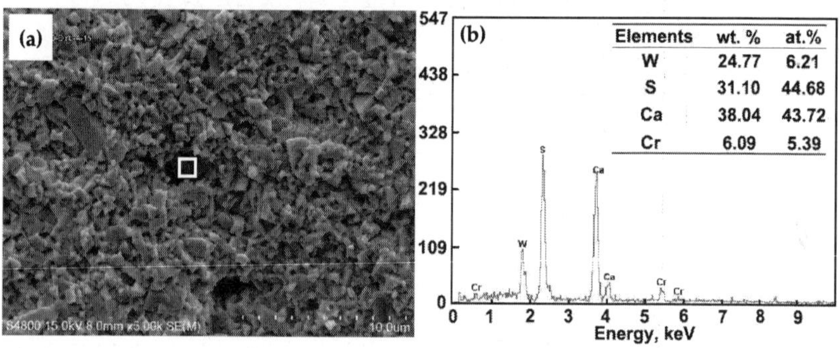

Figure 13. The failure of a sintered WC/Co alloy caused by the S and Ca impurities: (a) A fracture surface; (b) The EDS analysis at the fracture source (see the square region in (a))

4.3. Cobalt oxide

There are some advantages using Co_3O_4 powders to fabricate Co powders by hydrogen-reduction technology: (1) For $CoC_2O_4{\cdot}2H_2O$ and $CoCO_3{\cdot}xH_2O$ powders, Co powders easily grow if the decomposition and reduction processing (nucleation and growth processing) can not be completely separated. But for Co_3O_4 powders, there is only the reduction processing. So the technology processing is easily controlled. (2) The content of Co in Co_3O_4 is two times higher than those in $CoC_2O_4{\cdot}2H_2O$ and $CoCO_3{\cdot}xH_2O$, as listed in Tab. 4. So the output of Co powders can be improved by double times at a same reduction condition. (3) The Co powders obtained by hydrogen-reducing Co_3O_4 powders are almost all hcp-Co. Therefore, ultra-fine spherical Co_3O_4 powders are the best raw material to fabricate ultra-fine spherical Co powders.

Compounds	$CoC_2O_4{\cdot}2H_2O$	$CoCO_3{\cdot}xH_2O$	CoO	Co_2O_3	Co_3O_4
Cobalt contents (wt.%)	32.24	≤49.58	78.67	71.08	73.44

Table 4. Cobalt contents in various raw materials used to prepare cobalt powders

5. The continuous and controllable combustion synthesis

Basing on the above analysis, it is very difficulty to obtain the satisfied ultra-fine spherical Co powders by the conventional technique while using $CoC_2O_4{\cdot}2H_2O$ and $CoCO_3{\cdot}xH_2O$ as the raw materials. Ultra-fine spherical Co_3O_4 powders are the best raw material used to fabricate ultra-fine spherical Co powders. But the key question is how to obtain ultra-fine spherical Co_3O_4 powders by a low energy-consumption and low-cost method? And then a continuous and controllable combustion synthesis has been proposed.

5.1. Basic principles

5.1.1. Selecting the raw powders

As mentioned above, although spherical Co powders can be obtained by decomposing spherical $CoCO_3{\cdot}xH_2O$ powders, there is the higher content of some impurities (such as S and Ca) and the size of Co powders is still larger. So $CoC_2O_4{\cdot}2H_2O$ powders will be used as the raw materials to fabricate ultra-fine spherical Co_3O_4 powders, and then ultra-fine spherical Co powders can obtained by hydrogen-reduction technology in Fig.5. Using $CoC_2O_4{\cdot}2H_2O$ powders to fabricate ultra-fine spherical Co_3O_4 powders, the following advantages can be utilized and some key technical difficulties must be resolved.

5.1.2. Utilizing the exothermal reaction of $CoC_2O_4{\cdot}2H_2O$ — Low energy-consumption

Selecting $CoC_2O_4{\cdot}2H_2O$ powders as the raw material to fabricate ultra-fine Co_3O_4 powders, a main advantage is that the decomposition processing of $CoC_2O_4{\cdot}2H_2O$ in the air is an exothermal processing according to DTA. Seham *et al.* [23] showed that the total exothermal energy was 24.26kJ/mol during the decomposition of $CoC_2O_4{\cdot}2H_2O$ in the air, which is an enough energy to operate the following decomposition processing of $CoC_2O_4{\cdot}2H_2O$. Therefore, after giving a starting energy, the decomposition once starts, and then further heating does not need, and the needful energy of the following decomposition can be provided by the released reaction energy of CoC_2O_4 converting into Co_3O_4. So the decomposition processing of $CoC_2O_4{\cdot}2H_2O$ in the air is a spontaneous and continuous processing, and a lot of energy can be saved.

5.1.3. Utilizing airflow dispersion to in situ break the adhesion among Co_3O_4 particles

Morphology of Co_3O_4 powders can still inherit the short-fibrous structure of $CoC_2O_4{\cdot}2H_2O$ powders when the decomposition processing is carried out in a fixed bed, namely Co_3O_4 particles can easily grow together to form a short-fibrous structure (see Fig.14(a)) due to the exothermal reaction. But a short-fiber is composed of ultra-fine spherical Co_3O_4 particles with a size of about 0.1~0.2μm, see Fig.14(b). The Co_3O_4 particles arrange a short-string along to the fiber, and there is a weak adhesion between particles. The results further explain why the short-string or dendritic Co powders is easily obtained by decomposing and reducing $CoC_2O_4{\cdot}2H_2O$ powders in a fixed bed. Ultra-fine Co_3O_4 powders with the size of 0.1~0.2μm can obtained if the weak adhesion between them before forming a strong bond can be broken *in situ*. How to break the adhesion of Co_3O_4 particles? So a dynamic decomposing processing is proposed. If an airflow can be provided to $CoC_2O_4{\cdot}2H_2O$ powders during decomposition, the released energy can be homogenized to avoid overheat in part and also impede the adhesion and grow among Co_3O_4 particles.

Figure 14. The short-fibrous structure of Co_3O_4 particles obtained by decomposing $CoC_2O_4{\cdot}2H_2O$ powders in a fixed bed: (a) low multiple, the morphology with a short-fibrous structure is very similar to that of $CoC_2O_4{\cdot}2H_2O$ powders; (b) Magnifying several fibrous particles, consisted of ultra-fine spherical Co_3O_4 particles with a size of about 0.1~0.2μm

5.2. Production equipments

According to the above desires, a continuously dynamic-controlled combustion synthesis (CDCCS) process is proposed and the bottlenecks of continuous production and process controls of combustion synthesis (CS) have also been solved satisfactorily. The preparation of ultra-fine Co_3O_4 powders via CDCCS is carried out in a gas-solid fluidized bed unit (CS unit), as shown in Fig.15. The unit is patented equipment consisting of fluidized bed roaster, feeding systems, receiving systems, dust collection systems and air supply system [24,25].

There are reaction and cooling boiling (11 and 21) that formed by gas-solid fluidization on the upper and lower gas distribution plates (61 and 62), respectively. The raw material on the upper gas distribution plate (61) is fleetly penetrated by the gas from the upper gas pipeline (32). The boiling bed with the flowing property like liquid is formed when the superficial gas velocity reaches a critical value of 0.03m/s. In our experiments, the superficial gas velocity is 0.09~0.18m/s. The height of boiling beds, locating between the gas distribution plates to the top of overflow gates, is about 500mm. And several thermocouples are installed on the CS unit' wall and the location is about at 2/3 height of boiling beds.

When the boiling bed (11) containing $CoC_2O_4{\cdot}2H_2O$ powders which are fed into by the screw feeder (51) is preheated to 380°C by the hot carrier gas, $CoC_2O_4{\cdot}2H_2O$ powders will react with O_2 in the air. Co_3O_4 powders pour from the upper overflow pipe (35) under gravity and enter the cooling boiling bed (21), followed by collecting from the lower overflow pipe (34). The temperature of CS in reaction region can be perfectly controlled though the linkage of the upper thermocouple (71) and the upper gas pipeline (32).

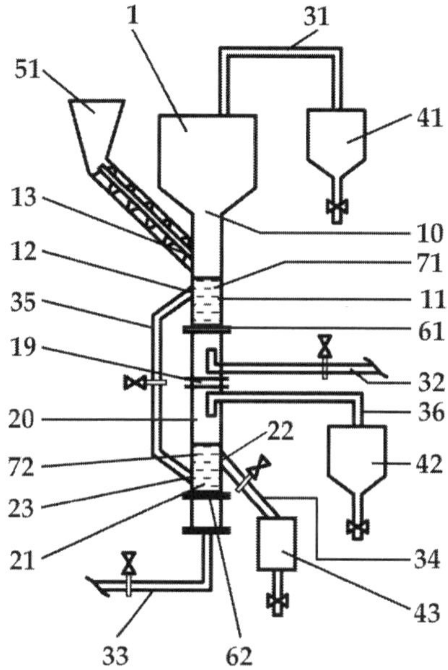

Figure 15. Schematic representation of CS unit, in which Co_3O_4 powders fabricated by CDCCS: 1. Fluidized bed roaster; 10. Upper chamber; 11. Reaction boiling bed; 12. Upper overflow gate; 13. Feed inlet; 19. Clapboard; 20. Lower chamber; 21 Cooling boiling bed; 22. lower overflow gate; 23. Overflow entry; 31. Upper exhaust pipe; 32. Upper gas pipeline; 33. Lower gas pipeline; 34. Lower overflow pipe; 35. Upper overflow pipe; 36. Lower exhaust pipe; 41. Upper dust catcher; 42. Lower dust catcher; 43. Product collection container; 51. Screw feeder with storage hopper; 61. Upper gas distributor; 62. Lower gas distributor; 71. Upper thermocouple; 72. Lower thermocouple

5.3. Process analysis of CDCCS to fabricate Co_3O_4 powders

When the reactant $CoC_2O_4{\cdot}2H_2O$ powders are fed by the screw feeder (51) into the boiling bed, no self continuous reaction occurs. Only when $CoC_2O_4{\cdot}2H_2O$ powders in the combustion zone are heated by hot carrier gas, and the continuous reaction is ignited. This means that once the combustion reaction between O_2 and $CoC_2O_4{\cdot}2H_2O$ powders is ignited, the released heat by the combustion reaction can be used to ignite the following exothermic reaction and no external heat is needed, and then the hot carrier can be transformed into the cool carrier gas.

Fig.16 shows the effect of the feeding rate of $CoC_2O_4{\cdot}2H_2O$ powders and the superficial gas velocities in the boiling bed on the temperature of combustion wave [26]. Temperature of combustion wave decreases with increasing gas velocity at a given feeding rate. Therefore, the temperature of combustion wave can be adjusted and controlled dynamically by changing the feeding rate of $CoC_2O_4{\cdot}2H_2O$ powders and/or the superficial gas velocity. According to the thermal analysis of $CoC_2O_4{\cdot}2H_2O$ in the air, the CoC_2O_4 can be completely converted to Co_3O_4 at a temperature higher than 320°C. Therefore, in order to ensure the dehydration reaction of $CoC_2O_4{\cdot}2H_2O$ and the oxidation reaction of CoC_2O_4 can be carried out rapidly and completely, and the combustion temperature is set at 380~400°C.

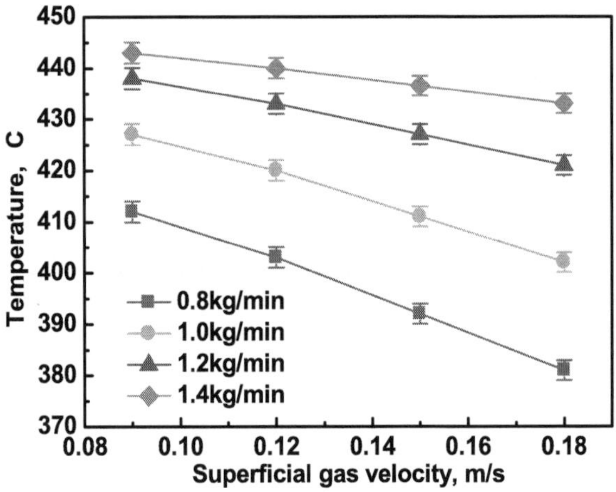

Figure 16. Effect of the feeding rate of $CoC_2O_4{\cdot}2H_2O$ powders and the superficial gas velocities in the boiling bed on the temperature of combustion wave

Being different from the conventional CS, the combustion wave here is almost full of the combustion zone in the middle of the boiling bed and the temperatures in whole space are homogeneous. The combustion wave is relatively suspended in the fluidized bed, and the temperature in the combustion zone can be dynamically controlled by adjusting the flow of carrier gas, as mentioned above. During the preparation, $CoC_2O_4{\cdot}2H_2O$ powders are continuously fed by the screw feeder into the upper area of reaction boiling bed and contacted with the combustion wave. The Co_3O_4 particles are dispersed by the carrier gas flow. After the CS, the dispersed Co_3O_4 powders fall down and leave the combustion wave in time through flowing out off the overflow gate on the wall of the unit continuously.

The CS unit in the paper is different from the conventional one reported in Ref. [27,28]. In the conventional unit, the reactants and products are nearly stationary while the combustion wave propagates from reactants to products. The advantages of the continuous fluidization technologies and the CS are well utilized and combined in our unit. The combustion wave is relatively stationary, but the reactants and products are continuous and mobile at a given speed.

5.4. Quality of products

Fig.17(a) shows the morphology of Co_3O_4 powders fabricated by the CDCCS. The adhesion among Co_3O_4 powders shown in Fig.14(b) is broken by the carrier gas flow. The particles exhibit a spherical or quasi-spherical shape, and the size of Co_3O_4 particles is about 0.1~0.3μm. In the gas-solid fluidized bed unit, the temperature of the whole boiling bed is nearly homogeneous, and almost each solid particle in the boiling bed has a similar surrounding, namely all $CoC_2O_4{\cdot}2H_2O$ particles can be completely decomposed to form Co_3O_4. The suspended solid particles collide and grind each other without aggregation by the action of airflow, which accelerates the formation of the spherical Co_3O_4 powders. The uneven distributions of particle size or hard aggregation appear scarcely due to the favorable diffusion condition and rapid transfer of heat and mass. The Co_3O_4 powders with a narrow particle size distribution, good dispersity and excellent fluidity are fabricated by the CDCCS.

As mentioned above, using the ultra-fine spherical Co_3O_4 powders fabricated by the CDCCS as raw materials, the ultra-fine spherical or quasi-spherical Co powders is easily obtained by hydrogen-reduction technology, as shown in Fig.17(b). The Co powders with an average size of 0.6μm possess a narrow particle size distribution, good dispersity, and excellent fluidity.

Figure 17. (a) Ultra-fine spherical Co_3O_4 powders fabricated by CDCCS; (b) Ultra-fine spherical Co powders fabricated by hydrogen-reduction technology using ultra-fine spherical Co_3O_4 powders as raw materials

5.5. Advantages of the CDCCS

The preparation of ultra-fine spherical Co_3O_4 powders (the size is smaller than 0.8μm and the length-diameter radio is smaller than 2) by the CDCCS has been successfully applied in the industrialization production in Xiamen Golden Egret Special Alloy Co., Ltd [24,25].The novel method has the following several advantages:

1. Comparison with the fixed bed, ultra-fine spherical Co_3O_4 powders are continuously produced in the fluidized bed unit and the production efficiency is improved.
2. No external heat is needed after the reaction is once ignited, and then the energy consumption is reduced.
3. The fabrication process is in a closed unit so that a clean operation environment is realized and a high pure product is fabricated.
4. The device operations are automated, except the transport of reactants and products, thus the labor intensity is largely decreased.
5. The most important advantage is that the properties of the products fabricated by CDCCS are more excellent than those fabricated by the conventional method.

6. Expectation

Ultra-fine spherical Co_3O_4 powders are firstly fabricated by CDCCS, and then ultra-fine spherical Co powders can be fabricated by hydrogen-reduction technology. Can ultra-fine spherical Co powders be directly fabricated by CDCCS if a hydrogen flow is blown into the CS unit? However, the hydrogen is a flammable and explosive gas. Therefore, our future work is how to directly fabricate ultra-fine spherical Co powders in the CS unit by solving the key technology difficulty of the hydrogen safety. Furthmore, the fabrication of many other metal powders (for example, W, Mo powders, etc.) can be applied in the CS unit.

Author details

Chong-Hu Wu
China National R&D Center for Tungsten Technology,
Xiamen Tungsten Co. Ltd. Technology Center, Xiamen, China;
Xiamen Golden Egret Special Alloy Co. Ltd., Xiamen, China

7. References

[1] Celozzi Salvatore, Araneo Rodolfo, Lovat Giampiero. Electromagnetic shielding. John Wiley & Sons, Hoboken, New Jersey, Canada, 2008, p. 27.

[2] Enghag Per. Cobalt. Encyclopedia of the elements: technical data, history, processing, applications. John Wiley & Sons, Hoboken, New Jersey, Canada, 2004, p. 667.

[3] V. S. R. Murthy, A. K. Jena, K. P. Gupta, G. S. Murty. Structure and properties of engineering materials. Tata McGraw-Hill Publishing Company Limited, 2003, p. 381.

[4] V. M. Schastlivtsev, Yu. V. Khlebnikova, T. I. Tabatchikova, D. P. Rodionov, V. A. Sazonova. Formation of a structure in cobalt single crystals at the $\beta \rightarrow \alpha$ transformation. Doklady Physics, 2009, 54(1): 21-24.

[5] B. W. Lee, R. Alsenz, A. Ignatiev, M. A. Van Hove. Surface structures of the two allotropic phases of cobalt. Physical Review B, 1978, 17(4): 1510-1520.

[6] Rong-jiu Li, Hong-qiang Jia, Xu-dong Sun. Ceramics-metal composites. Metallurgical Industry Press, Beijing, China, 2002, p. 233-239. [in Chinese]

[7] Chen Biao. Study on world cobalt mineral resources and technology of cobalt extraction in difficultly extracting cobalt ore. Doctoral Dissertation, Jilin University, Jilin, China, 2001, p. 27-29. [in Chinese]

[8] http://en.wikipedia.org/wiki/Cobalt.

[9] http://www.indexmundi.com/en/commodities/minerals/cobalt/cobalt_t10.html

[10] Xing-long Tan, Mao-zhong Yi, Chong-ling Luo. Preparation of spherical cobalt powder and its application in ultra-fine cemented carbides. The Chinese Journal of Nonferrous Metals, 2008, 18(2): 209-214. [in Chinese]

[11] M. H. Tikkanen, A. Taskinen, P. Taskinen. Characteristic properties of cobalt powder suitable for hard metal production. Powder Metallurgy, 1975, 18: 259-282.

[12] Li-wei Huang. Study on the decomposition mechanism of cobalt oxalate. Nonferrous Metals (Extractive Metallurgy), 2005, 3: 40-43. [in Chinese]

[13] Jin Gao, Hong-jun Wang. Influence of morphology of precursor particles on Co powder morphology. Rare Metals and Cemented Carbides, 2002, 30(2): 15-18, 27. [in Chinese]

[14] Jin Gao, Qing-lin Chen. Influence of active spheroidization agent on the morphology of cobalt oxalate. Rare Metals and Cemented Carbides, 2001, 146: 20-22. [in Chinese]

[15] Hui-ling Du, Jian-zhong Wang, Jin-gang Qi. Effects of pulsed electromagnetic field on $CoC_2O_4{\cdot}2H_2O$ powder size. Acta Metallurgica Sinica, 2009, 45(8): 1019-1024. [in Chinese]

[16] Li-wei Huang. Study on the effecting factors on cobalt oxalate decomposition and the size of cobalt powder. Nonferrous Metals (Extractive Metallurgy), 2007, 1: 41-45. [in Chinese]

[17] Wei Li, Qin-sheng Zhao. Research on characteristic properties of tungsten-doped cobalt powder. Cemented Carbide, 1997, 14(4): 204-206. [in Chinese]

[18] Yi-yong Yang, Ju-tang Sun, Liang-jie Yuan, Ke-li Zhang. Micro-method powder X-ray diffraction analysis of thermal decomposition product of basic cobalt carbonate. Journal of Wuhan University, 2001, 47(6): 660-662. [in Chinese]

[19] Xing-long Tan, Mao-zhong Yi, Chong-ling Luo. Preparation of spherical cobalt powder and its application in ultra-fine cemented carbides. The Chinese Journal of Nonferrous Metals, 2008, 18(2): 209-214. [in Chinese]

[20] Chong-ling Luo, Mao-zhong Yi, Xing-long Tan. Discussion on the technology for producing ultra-fine spherical cobalt powder from carbonate. Cemented Carbide, 2007, 24(2): 84-87. [in Chinese]

[21] J. Xiong, J. G. Yang, X. H. Guo. Application of rare earth elements in cemented carbide inserts, drawing dies and mining tools. Materials Science and Engineering A, 1996, 209: 287-293.

[22] C. H. Xu, X. Ai, C. Z. Huang. Research and development of rare-earth cemented carbides. International Journal of Refractory Metals and Hard Materials, 2001, 19: 159-168.

[23] A. S. Mansour, Spectrothermal studies on the decomposition course of cobalt oxysalts. Part III. Cobalt oxalate dehydrate. Materials Chemistry and Physics, 1994, 36(3-4): 324-331.

[24] Chong-hu Wu. Energy-saving method and device for continuous production of tricobalt tetraoxide by cobaltous oxalate. CN 101062791, 2007.

[25] Chong-hu Wu. Energy saving type device for continuously producing Co_3O_4 by cobalt oxalate. CN 2900494, 2007.

[26] Chong-hu Wu. Preparation of ultrafine Co_3O_4 powders by continuous and controllable combustion synthesis. Transactions of Nonferrous Metals Society of China, 2011, 21: 679-684.

[27] Zhi-jie Zhang, Wen-zhong Wang, Meng Shang, Wen-zong Yin. Low-temperature combustion synthesis of Bi_2WO_6 nanoparticles as a visible-light-driven photocatalyst. Journal of Hazardous Materials, 2010, 177(1-3): 1013-1018.

[28] E. Magnone, E. Traversa, M. Miyayama. Nano-sized $Pr_{0.8}Sr_{0.2}Co_{1-x}Fe_xO_3$ powders prepared by single-step combustion synthesis for solid oxide fuel cell cathodes. Journal of Electroceramics, 2010, 24(2): 122-135.

Chapter 7

Production and Characterization of Biofuel from Non-Edible Oils: An Alternative Energy Sources to Petrol Diesel

A.S. Abdulkareem, A. Jimoh, A.S. Afolabi, J.O. Odigure and D. Patience

Additional information is available at the end of the chapter

http://dx.doi.org/10.5772/51341

1. Introduction

The production and consumption of petroleum oil increases constantly and it has been reported that about 75 million barrels of crude oil is consumed worldwide daily [1-4]. Resulting energy crisis as a result of over dependence on fossil fuel as a source of energy have shown that fossil fuels are limited finite resource and the continuous depreciation of the world oil reserves also corroborates the fact that oil is a finite non-renewable source of energy that will ultimately be exhausted [5-9]. This therefore leads to continuous agitation and research activities towards establishing alternative renewable fuels that can replace and prevent possible negative impacts that may result from continuous use and depreciation of fossil fuel [10]. The need to reduce complete dependence on fossil fuel as a sole energy sources and need to source for alternative energy cannot be over emphasized. Presently, biofuel, biogas and bio ethanol are considered as the most promising alternatives in energy generation that can compete with the fossil fuel [2, 11-12]. Biodiesel is produced in some countries and used efficiently either alone or in blends with minerals diesel in cars and transport vehicles. The transesterification of the triglycerides present in plant oils to the methyl or ethyl esters. Since the passage of the Energy Policy Act of 2005, biodiesel use has been increasing especially in the developed countries. For instance, biodiesel is produced in some countries and used efficiently either alone or in blends with mineral diesel in cars and transport vehicles [13]. The United State Fueling stations make biodiesel readily available to consumers across Europe and increasingly in the USA and Canada. This is an indication that biodiesel can operates in compression ignition engines like petroleum diesel without requiring no essential engine modifications [14-16]. Moreover it can maintain the payload capacity and range of conventional diesel unlike fossil diesel, pure biodiesel is biodegradable, nontoxic and essentially free of sulphur and aromatics [17-19]. A growing number of transport fleets use it as an additive in their fuel. Biodiesel is being used in both public and private fleet vehicles due to environmentally friendly and the facts that it offers a reduction in some emissions without requiring any modifications to the vehicle of biodiesel [14]. It is worth of mentioning that the production of this alternative energy source (biodiesel), which is a variety of ester-based oxygenated fuels, can be produced from different feedstock, such as vegetable oil, animal fats and non edible oils [20-24]. Hence, the production of biodiesel is sensitive to feedstock; care must therefore take in selecting the feed stock for the production of biofuel [14]. For instance production of biofuel from vegetable oils had been criticized by the experts as a source of food crisis [14]. The critics of this method of biofuel production beliefs that production of biofuel from consumable oil will resulted into food crisis and price instability. Also militate against the commercial availability of biodiesel is the cost of biodiesel. For instance, biodiesel costs over US$0.5/l compared to US$0.3/l for petrol diesel, the is due mainly to the cost of virgin vegetable oil used as a feedstock for the production of biodiesel [14]. For instance, in the United State, the cost of soybean is on the average of US$0.36/l, Currently, the cost of biodiesel stands at 3.83 dollars per gallon which is 19.6% higher compared to that of petro-diesel ($ 3.07) [25]. It is therefore not surprising that the biodiesel produced from pure soybean is much more expensive than petroleum based diesel. Hence, the recent research and development in the production of biofuel as alternative energy sources that will compete well with the existing energy sources (fossil fuel) is now concentrated on in-edible oil, which is the focus of this present study [26-31]. This work involves comparative study of biodiesel production from Jatropah curcas and Ricininus communis. The choice of Jatropha curcas and Ricinus communis as a feedstock for the production of biodiesel is also favoured by the abundantly availability of these plants in Nigeria, despite their abundant availability, they are not properly exploited for technological and material development. Hence, Production of biodiesel from these oils will lead to proper utilization of these vegetable oils. It also includes characterization of both the oils and biodiesel produced. The focus of this study can be achieved through the following objectives;

- Characterization of both jatropha curcas and ricinus commumis oils to determine density, viscosity, saponification number, iodine value, acid value, peroxide value and moisture content which are compared with the standard.
- Production of biodiesel from jatropha curcas and ricinus commumis by two-step transesterification reaction.
- Investigation into the effect of temperature and catalyst concentration on the production of biodiesel from oil.
- Characterization of the biodiesel produced to determine the density, flash point, Kinematic viscosity, Cloud point, Sulphur content, Bottom Sediment & Water, Cetane number, water by distillation and distillation properties and compare with that of convention diesel fuel and available standard.
- Comparison of the qualities of biodiesel produced from Jatropha curcas and Ricinus communis with that of the standard values of biodiesel

1.1. History of biodisel

The name 'biodiesel' has been given to transesterified vegetable oil to describe its use as a diesel fuel. Transesterification of triglycerides in oils is not a new process. Scientists E. Duffy and J. Patrick conducted it as early as 1853 [11, 32]. Rudolf Diesel's prime model, a single 10 ft (3 m) iron cylinder with a flywheel at its base, ran on its own power for the first time in Augsburg, Germany, on 10 August 1893 running on nothing but peanut oil. In remembrance of this event, 10 August has been declared "International Biodiesel Day". Life for the diesel engine began in 1893 when the famous German inventor Rudolph Diesel published a paper entitled 'the theory and construction of a rational heat engine'. What the paper described was a revolutionary engine in which air would be compressed by a piston to a very high pressure thereby causing a high temperature. Rudolph Diesel designed the original diesel engine to run on vegetable oil [14,32]. Dr Rudolph Diesel used peanut oil to fuel one of his engines at the Paris Exposition of 1900. The engine was constructed for using mineral oil, and was then worked on vegetable oil without any alterations being made. The French Government at the time thought of testing the applicability to power production of the Arachide, or earth-nut, which grows in considerable quantities in their African colonies, and can easily be cultivated there." Diesel himself later conducted related tests and appeared supportive of the idea [32]. Due to the high temperatures created, the engine was able to run a variety of vegetable oil including hemp and peanut oil. At the 1911 World's Fair in Paris, Dr R. Diesel ran his engine on peanut oil and declared 'the diesel engine can be fed with vegetable oils and will help considerably in the development of the agriculture of the countries which use it.' One of the first uses of transesterified vegetable oil was powering heavy-duty vehicles in South Africa before World War II [32]. In a 1912 speech Diesel said, "The use of vegetable oils for engine fuels may seem insignificant today but such oils may become, in the course of time, as important as petroleum and the coal-tar products of the present time." Despite the widespread use of fossil petroleum-derived diesel fuels, interest in vegetable oils as fuels for internal combustion engines was reported in several countries during the 1920s and 1930s and later during World War II. Belgium, France, Italy, the United Kingdom, Portugal, Germany, Brazil, Argentina, Japan and China were reported to have tested and used vegetable oils as diesel fuels during this time [32]. Some operational problems were reported due to the high viscosity of vegetable oils compared to petroleum diesel fuel, which results in poor atomization of the fuel in the fuel spray and often leads to deposits and coking of the injectors, combustion chamber and valves. Attempts to overcome these problems included heating of the vegetable oil, blending it with petroleum-derived diesel fuel or ethanol, pyrolysis and cracking of the oils [14]. Research into the use of transesterified sunflower oil, and refining it to diesel fuel standards, was initiated in South Africa in 1979 [32]. By 1983, the process for producing fuel-quality, engine-tested biodiesel was completed and published internationally. An Austrian company, Gaskoks, obtained the technology from the South African Agricultural Engineers; the company erected the first biodiesel pilot plant in November 1987, and the first industrial-scale plant in April 1989 (with a capacity of 30,000 tons of rapeseed per annum) [32].

Throughout the 1990s, plants were opened in many European countries, including the Czech Republic, Germany and Sweden. France launched local production of biodiesel fuel (referred to as diester) from rapeseed oil, which is mixed into regular diesel fuel at a level of 5%, and into the diesel fuel used by some captive fleets (e.g. public transportation) at a level of 30%. Renault, Peugeot and other manufacturers have certified truck engines for use with up to that level of partial biodiesel; experiments with 50% biodiesel are underway. During the same period, nations in other parts of the world also saw local production of biodiesel starting up: by 1998, the Austrian Biofuels Institute had identified 21 countries with commercial biodiesel projects. 100% Biodiesel is now available at many normal service stations across Europe [32]. In September 2005 Minnesota became the first U.S. state to mandate that all diesel fuel sold in the state contain part biodiesel, requiring a content of at least 2% biodiesel [32]. Today's diesel engines require a clean burning and stable fuel that performs well under the variety of operating condition. Biodiesel is the only alternative fuel that can be used directly in any existing unmodified diesel engine. It has been reported that biodiesel can be blended with the diesel fuel in any proportion due to the similarities in their properties and the current trends in the production of biodiesel from vegetable are now concentrated on non- edible oil as the feedstock. Vegetable oils such as soybean, rapeseed and peanut are edible and expensive though the seeds of these oils are available in large quantities especially in Africa (Tamil Nadu -641 003india) [32] .However, there is need to control their utilization for the production of biodiesel for the purpose of not causing large problem of hunger and deforestation while sourcing for alternative energy sources [33-35].

1.2. Biodiesel

Biodiesel is defined as the mono-alkyl esters of fatty acids derived from a renewable lipid feedstock such as vegetable oils or animal fats [2, 16, 36-37]. In simple terms, biodiesel is the product obtained when a vegetable oil or animal fat is chemically reacted with an alcohol to produce fatty acid alkyl esters. A Catalyst such as sodium or potassium hydroxide is required. Glycerol is produced as a co product. Vegetable oils are a renewable and potentially inexhaustible source of energy with a calorific value close to that of conventional diesel fuel. Their high viscosity, low volatility, incomplete combustion and possibility of formation of deposits are the drawbacks of biodiesel. Biodiesel besides not being refined from crude oil, biodiesel offers other attractive characteristics, such as a significantly lower emission of carbon monoxide, CO (46.7 %,) carbon dioxide, CO_2, (78%), [38], sulphur dioxide, SO_2 and hydrocarbons, HC (45.2%) [39]. It also eliminates the cloud of dense, black smoke normally associated with diesel vehicles and other particulate matter (PM) emissions (66.7%) that cause respiratory damage to a level in line with the Kyoto Protocol Agreement [40]. Biodiesel is also reported to be highly biodegradable in freshwater as well as soil environments. 90–98% of biodiesel is mineralized in 21–28 days under aerobic as well as anaerobic conditions in comparison to 50% and 56% by diesel fuel and gasoline respectively [41]. European tests of rape seed oil-based biodiesel indicate that it is 99.6% biodegradable within 21 days [42]. Thus within one month of being spilled into the environment, biodiesel should be completely decomposed. This makes biodiesel a good alternative fuel for use in environmentally sensitive areas where fuel leakages and spills would be particularly harmful such as wetland areas and watersheds which supply drinking water. It has been reported also that biodiesels have high cetane number, high flash point and acceptable level of sulphur content which makes it very attractive as an alternative fuel [43]. Reported work by Korbitz (1999) [44] shows that the cetane number of biodiesel fuels is higher than fossil diesel which is considered to be significantly advantageous in terms of engine performance. The flash point of biodiesel fuels is twice as much as that of fossil diesel. Therefore, biodiesel is much safer to store, handle and use than the conventional diesel fuel [45]. Biodiesel fuels have been praised as being sulphur-free and may contain only traces of sulphur emanated from the acidic catalyst used. Also favour the choice of biodiesel as alternative sources of energy is the fact that biodiesel is virtually compatible with commercial diesel engines and practically no engine modifications are required [46]. Biodiesel can also be blended at any level with petroleum diesel to create a biodiesel blend. When blended with fossil diesel, it shows positive synergic effect of biodegradation by means of co-metabolism. For example, the time taken to reach 50% biodegradation of fossil diesel is reduced from 28 to 22 days in 5% biodiesel mixture [47]. In terms of economic benefit, biodiesel production is expected to remain profitable and grow dramatically. Farmers that produce highly rated energy crops such as Jatropha for the production of biodiesel stand a chance of improving their income level by providing market outlets for farmers and generate rural employment. The by-product of transesterification reaction, glycerol, when purified can be used in its traditional application (pharmaceutical, cosmetics and food industries). In addition, the obtained glycerol can be used in recently developed applications in the fields of animal feed, carbon feedstock in fermentations, polymers, surfactants and lubricants [48]. Biodiesel, thus, increases the contribution of agriculture to exports and economic growth. It will also generate new industries which will create new jobs and new markets for the majority of unemployed youth [49].

Despite the advantages of biodiesel as perfect alternative energy sources, emission of NOx is one of the setbacks of biodiesel. The temperature within the cylinders of a vehicle fuelled with biodiesel would increase due to the enhanced combustion as a result of high oxygen content of biodiesel. This increase in temperature stimulates the production of NOx from the reaction with nitrogen in the air, which results in a small increase in NOx emission compared to those produced from conventional diesel fuel [50]. Aside from the formation of NOx by the engine powered with biodiesel, the chemical contents of biodiesel is also a fatty acid methyl ester when the alcohol used during transesterification is methanol or fatty acid ethyl ester in case of ethanol. These ester molecules are susceptible to hydrolytic and oxidization reactions resulting in the formation of polymers. This makes the biodiesel unstable on storage and hence cannot sit on the shelf for long time as it develops mould when it gets old [51]. Exploring the means of producing biodiesel that will compete well with the existing petroleum diesel is of much interest in the recent biodiesel research, especially for those methods concentrating on minimizing the raw material cost. Biodiesel as alternative energy source can be produced from different sources such as soybean, beef tallow, waste vegetable oil etc depends on the location and availability. The fact that biodiesel burns clearly and source from renewable sources widens its acceptance world wide as alternative energy source. Biodiesel which is described as variety of ester-based oxygenated fuels derived from natural, renewable sources such is a perfect alternative energy sources due to its biodegradability nature. To improve the quality of vegetable oil, researchers have employed different methods listed below to reduce the high viscosity of the oils.

1. Dilution of 25 parts of vegetable oil with 75 parts of diesel fuel (biodiesel blends)
2. Micro emulsions with short chain alcohol (e.g. ethanol or methanol).
3. Thermal decomposition which produces alkanes, alkenes, carboxylic acids and aromatic compounds.
4. Catalytic cracking which produces alkanes, cycloalkanes and alkyl benzene.
5. Transesterification.

Transesterification is described as the best choice as it is relatively simple and produces a product with properties close to diesel fuel. Alcohol, oil and catalyst are combined in an agitated reactor, approx. at 60ºC during 1h. Smaller plants often use batch mode reactors, but larger plants use continuous flows processes involving continuous stirred-tank reactors or plug flow reactors. The processes involved in biodiesel production from feedstock containing low levels of FFA which, include soybean oil, canola (rapeseed) oil, and the higher grades of waste restaurant oils. Alcohol, catalyst, and oil are combined in a reactor and agitated for 1hour at 60°C. Smaller plants often use batch reactors [32] but most larger plants (4 million L/yr) use continuous flow processes involving continuous stirred-tank reactors (CSTR) or plug flow reactors [32]. The reaction is sometimes done in two steps in which 80% of the alcohol and catalyst is added to the oil in a first-stage CSTR. Then, the product stream from this reactor goes through a glycerol removal step before entering a second CSTR. The remaining 20% of the alcohol and catalyst is added in this second reactor. This system provides a very complete reaction with the potential of using less alcohol than single-step systems. After the reaction, glycerol is removed from the methyl esters. Due to the low solubility of glycerol in the esters, this separation generally occurs quickly and can be accomplished with either a settling tank or a centrifuge. The excess methanol tends to act as a solubilizer and can slow the separation. However, this excess methanol is usually not removed from the reaction stream until after the glycerol and methyl esters are separated due to concern about reversing the transesterification reaction. Water may be added to the reaction mixture after the transesterification is complete to improve the separation of glycerol [32].After separation from the glycerol, the methyl ester enter a neutralization step and then pass through a methanol stripper, usually a vacuum flash process or a falling film evaporator, before water washing. Acid is added to the biodiesel product to neutralize any residual catalyst and to split any soap that may have formed during the reaction. Soaps will then react with the acid to form water-soluble salts and FFA. The salts will be removed during the water washing step and the FFA will stay in the biodiesel. The water washing step is intended to remove any remaining catalyst, soap, salts, methanol, or free glycerol from the biodiesel. Neutralization before washing reduces the amount of water required and minimizes the potential for emulsions to form when the wash water is added to the biodiesel. After the wash process, any remaining water is removed from the biodiesel by a vacuum flash process. The glycerol stream leaving the separator is only 50% glycerol. It contains some of the excess methanol and most of the catalyst and soap. In this form, the glycerol has little value and disposal may be difficult. The methanol content requires the glycerol to be treated as hazardous waste. The first step in refining the glycerol is usually to add acid to split the soaps into FFA and salts. The FFA is not soluble in the glycerol and will rise to the top where they can be removed and recycled[32]. Mittelbach and Koncar [32] described a process for esterifying these FFA and then returning them to the transesterification reaction stream. The salts remain with the glycerol, although depending on the chemical compounds present, some may precipitate out. One frequently touted option is to use potassium hydroxide as the reaction catalyst and phosphoric acid for neutralization so that the salt formed is potassium phosphate, which can be used for fertilizer. After acidulation and separation of the FFA, the methanol in the glycerol is removed by a vacuum flash process, or another type of evaporator. At this point, the glycerol should have a purity of 85% and is typically sold to a glycerol refiner. The glycerol refining process takes the purity up to 99.5–99.7% using vacuum distillation or ion exchange processes. Therefore, biodiesel is defined as the mono-alkyl esters of fatty acids derived from vegetable oils, animal fats by transesterification with alcohol in the presence of catalyst. A catalyst such as sodium or potassium hydroxide is required; glycerol is produced as a by-product.

2. Jatropha curcas and ricininus communis (castor oil) and biodiesel production

2.1. Jatropha curcas

Jatropha is a drought resistance tree that can be grown in areas where rainfall is as low as 500-600mm. It does well in Cape Verde where rainfall is only 250mm but the air is humid. In dry season, it tends to shed its leaves. Where rainfall is high above 1000mm, it does better in hot rather than temperate climate. Jatropha grow in soils that are quite infertile. It is usually found at lower elevations (below 500m) (14,32]. Jatropha Curcas is mostly found in northern part of Nigeria e.g. Sokoto, Zamfara, Kebbi, Kastina Minna etc [32].

There is growing interest for biodiesel production from non-edible oil source, like Jatropha curcas . Jatropha curcas is a plant belonging to Euphorbiaceae family, which is a non-edible oil-bearing plant widespread in arid, semi-arid and tropical regions of the world[32]. Jatropha curcas has an estimated annual production potential of 200 thousand metric tons in India and can grow in waste land [32]. Jatropha curcas is native to Mesoamerica but has been distributed worldwide since the XVI century by Spaniards and Portuguese to their tropical and subtropical colonies in Africa and Asia. In India and other countries of the Far East, villagers used Jatropha curcas as a hedge crop, and the extracted seed oil to make soap or fuel for lamps. In Europe, biodiesel production is based on the methyl ester of rapeseed oil, while in the United States, the methyl esters of soybean and rapeseed oils are more used. In tropical countries, Jatropha curcas seed oil is being promoted for biodiesel production, and the technology has been optimized [51]. It was found that the shell could be for combustion, hull/husk for gasification, and cake for production of biogas, spent slurry as manure, oil and biodiesel (made from Jatropha oil) for running CI engines. The kernels) of jatropha curcas have about 50% oil. The oil recovery in mechanical expeller was about 85%, while more than 95% recovery of oil could be achieved when extracted by solvent method. The biodiesel from jatropha curcas oil has a great potential due to its comparable properties to diesel, such as calorific value and cetane number [51]. Therefore, many researchers have shown great interest in using Jatropha oil to produce biodiesel. Azam *et al.* (2005) [52] found FAME of Jatropha curcas were most suitable for use as biodiesel and it met the major specification of biodiesel standards of USA, Germany and European Standard Organization. Sarin *et al.* (2007) [51] made 25 appropriate blends of Jatropha and palm biodiesel to improve oxidation stability and low temperature property based on the fact that Jatropha biodiesel has good low temperature property and palm biodiesel has good oxidative stability. It was found that antioxidant dosage could be reduced by 80-90% when palm oil biodiesel is blended with Jatropha biodiesel at about 20-40%. This techno-economic combination could be an optimum mix for Asian Energy Security. Tiwari *et al.* (2007) [53] used response surface methodology to optimize three important reaction variables, including methanol quantity, acid concentration, and reaction time. The optimum combination for reducing the FFA of Jatropha oil from 14% to less than 1% was found to be 1.43% v/v sulfuric acid catalyst, 0.28 v/v methanol-to-oil ration and 88 min reaction time at 60ºC for producing biodiesel. The properties of Jatropha oil biodiesel conformed to the American and European standards. As comparison, Berchmans *et al.* (2008) [54] developed a two-step pretreatment process in which the high FFA (15%) of Jatropha curcas seed oil was reduced to less than 1%. In the first step, the reaction was carried out with 0.60 w/w methanol-to-oil ratio in the presence of 1 wt.% sulfuric acid as an acid catalyst in 1h at 50ºC. In the second step, the transesterification reaction was performed using 0.24 w/w methanol-to-oil ratio and 1.4 wt.% sodium hydroxide as alkaline catalyst to produce biodiesel at 65ºC. The final biodiesel yield of 90% in 2hours was reported. As well as experimental study, theoretical studies of reaction mechanism were also conducted regarding to base-catalyzed transesterification of the glycerides of the *Jatropha* oil (Tapanes, Aranda, Carneiro & Antunes, 2008). In that study, semi-empirical AM1 molecular orbital calculations were used to investigate the reaction pathways of base, catalyzed transesterification of glycerides of palmitic, oleic and linoleic acid. The researchers concluded that the reaction mechanism included three steps: Step 1- Nucleophilic attack of the alkoxide anion on the carbonyl group of the glyceride to form a tetrahedral intermediate. Step 2-Breaking of the tetrahedral intermediate to form the alkyl ester and the glyceride anion. Step 3-Regeneration of the active catalyst, which may start another catalytic cycle. This study suggested that the Step 2, decomposition of the tetrahedral intermediate, determined the rate of base-catalyzed transesterification of glycerides. A lot of different approaches were taken when producing biodiesel from *Jatropha* oil. In additional to conventional methods, preparation of biodiesel from *Jatropha* oil using ultrasonic energy was investigated (Kachhwaha, Maji, Faran, Gupta, Ramchandran & Kumar, 2006). Low frequency ultrasound (33 kHz) was applied to transesterify Jatropha *oil* with methanol in the presence of base catalyst at 6:1 methanol/oil molar ratio. The reaction time (about 15-30 min) was much shorter than conventional mechanical stirring method. This method was proved to be efficient and economically functional. Moreover, enzyme catalysts were also utilized for biodiesel production from Jatropha *oil*. Shah *et al*. *Jatropha* oil in presence of supercritical carbon dioxide. The optimum conditions were found to be 8h, 45ºC, 5:1 molar ratio of alcohol to oil and an enzyme loading of 30% based on the weight of oil. However, conversions of only 60-70% were obtained even after 8 hours. The authors attributed this to the fact that the enzymatic reaction encountered both substrate and product inhibition. In contrast, when synthesis of biodiesel in supercritical alcohols, high conversions (80%) were obtained within 10 min and nearly complete conversions were obtained within 40 min. Despite of expected high operating cost due to high temperature and pressure associated with supercritical alcohol, it was still considered to be economically feasible since the reaction time was very short [32]. Furthermore, the absence of pre-treatment step, soap removal, and catalyst removal can significantly reduce the capital cost of a biodiesel plant. Meanwhile, many researches were conducted aiming at evaluating the performance, emission, and combustion characteristics in a diesel engine for Jatropha oil and Jatropha oil biodiesel [32] .Haldar *et al*. (2008) [55] found that Jatropha oil gave the best results related to the performance and emissions, such as CO, CO_2, HC, smoke and particulates, at high loads and 45º before Top Dead Center (BTDC) injection timing when compared with non-edible straight vegetable oils of Putranjiva, Jatropha and Karanja. Though received a booming interest due to its general characteristics and potentiality was recommended by some researchers that better data are urgently needed to guide investment since uncertainty do exist, based on the fact that Jatropha curcas *is* still a wild plant which exhibits a lot of variability in yield, oil content and oil quality [32].

2.2. Ricininus communis (castor oil)

Castor oil is possibly the plant oil industry's most underappreciated asset. It is one of the most versatile of plant oils, being used in over ten diverse industries. Owing to its unique chemical structure, castor oil can be used as the starting material for producing a wide range of end-products such as biodiesel, lubricants and greases, coatings, personal care and detergent, surfactants, oleo chemicals e.t.c. Compared to many other crops, castor crop requires relatively fewer inputs such as water, fertilizers and pesticides. The crop can also be grown on marginal land, thus providing an excellent opportunity for many regions of the world to utilize their land resources more productively [32]. The plant prefers well-drained moisture relative clay or sandy loan in full sun requires a rich soil and day time temperature above 20°C for seedling to grow well. Castor is native to tropical Africa but it grows widely in Nigeria as weed it can be found in Borno, Sokoto, Jos, Zaria and so many other places in the country [32]. Though, it has been reported that the plants is not properly exploited, however a fiber for making ropes can be obtained from its stem. The growing plant is said to repel flies and mosquitoes when grown in the garden and it is also said to rid it of moles and nibbling insect, while the leaves have insecticidal properties.

Cultivation of castor plants for the production of biodiesel started in 2008 in the Waletia and Goma Gofa regions of Ethiopia, the initiative is run by energy company (Global energy Ethiopia) who are also conducting research and development program to create new varieties of castor with better yields(sub-sahara Africa gateway, science and development network, 2008). Castor oil contains a number of fatty acids similar to those in cooking oils such as oleic, palmitic acids e.t.c. However, among vegetable oils castor oil is distinguished by its high content of ricinoleic acid (about 89.5 %). Castor oil unsaturated bond, molecular weight (298), low melting point (5 °C) makes it industrially useful. Castor oil is unique among all fats and oils, it has an unusual composition of a triglyceride of fatty acids, and it is the only source of an 18-carbon hydroxylated fatty acid with one double bond (Aldrich, 2003). The chemical composition of castor oil is given below [32]:

- Ricinoleic acid - 89.5 %
- Linoleic acid - 4.2 %
- Oleic acid - 3 %
- Stearic acid - 1 %
- Palmitic acid - 1 %
- Dihydroxystearic acid - 0.7 %
- Exosanoic acid - 0.3 %

Biodiesel derived from castor oil rates high among other oils with ash content of about 0.02 %, sulfur content less than 0.04 %, negligible potassium content, 35 GJ/T which compares favorably with other vegetable oils and petro diesel of about 45 GJ/T, and viscosity that is much higher than petro diesel, but this major bottleneck of the viscosity can be considerably reduced by transesterification making it a very efficient source of biodiesel

3. Material and methodology

3.1. Material and equipment

This study focus on the conversion of oil from Jatropha Curcas and Ricininus communis (castot oil) into biodesel by means of eseterification. The entire chemicals used in this study are of analytical grade (98-99.5%). They include Carbon tetra chloride (Analar, BDH), Wij's solution (Mixture of glacial acetate, iodine trichloride and carbon tetralchloride) (Hopkins and Williams, London), potasium hydroxide solution (Analar, BDH), potassium hydroxide pellet (Burgoyne & co, India), petroleum ether (Analar, BDH), potassium iodide solution (M&B, England), sodium thisosulphate (M&B, England), hydrochloric acid (Analar, BDH) and potassium iodide pellet (M&B, England). The equipments used are pH meter, distillation apparatus, viscometer, thermostatic hot plate, sulphur in oil analyzer, magnetic stirrer, digital weighing balance, thermometer, petri dish, pippete, separating funnel, burete, oil test centrifuge, pycometer bottles, Abbe refractometer, flash point tester, oven and aneline point teseter.

3.2. Methodology

3.2.1. Physical and chemical characterization of the oil

Prior to the production of biodiesel from jatropha cacus and risininus communis (castor oil), various anlysis such as specific gravity, acid value, free fatty acid, saponification value, iodine value, peroxide value, viscosity, moisture content and refarctive index were conducted on the oil samples.

3.2.1.1. Determination of specific gravity

A 50ml pycometer was washed thoroughly with detergent, water and petroleum ether, it was then dried and weighed. The bottle was filled with water and weighed, the bottle was then dried and filled with the oil sample and weighed, from theory, and the density of a substance was equal to mass of a substance per unit volume of that substance.

3.2.1.2. Determination of Acid Value/Free Fatty Acid (FFA)

2g of the oil was measured and poured in a beaker. A neutral solvent (a mixture of petroleum ether and ethanol) was prepared and 50ml of it was taken and poured into the beaker containing the oil sample. The mixture was stirred vigorously for 30minutes. 0.56g of potassium hydroxide (KOH) pellet was measured and placed in a separate beaker and 0.1M KOH was prepared, 3drops of phenolphthalein indicator was added to the sample and was titrated against 0.1M KOH till the color change observed turned pink and persisted for 15minutes.

$$\text{AV} = \frac{56.1 \times V \times N}{W_{OIL}} \quad (1)$$

Where; V= volume of standard alkali used; N= normality of standard alkali used; W_{oil}= weight of oil used

$$\text{FFA} = \frac{AN}{2} \quad (2)$$

3.2.1.3. Determination of saponification value

The alcoholic KOH was freshly prepared by dissolving KOH pellet in ethanol. 2g of oil was measured and poured into a conical flask. 25ml of the alcoholic KOH was added to it, a blank was used. The sample was well covered and placed in a steam water bath for 30minutes shaking it periodically, 1ml of phenolphthalein was added to the mixture and titrated against 0.5M HCl to get the end point.

$$SV = \frac{5.61 \times (B-A) \times N}{W_{OIL}} \quad (3)$$

Where; B= volume of standard ethanol potassium hydroxide used in blank titration; A= volume of standard ethanol potassium hydroxide used in titration with the oil; N= normality of standard acid; W_{oil}= weight of oil used.

3.2.1.4. Determination of iodine value

The oil was poured into a small beaker, a small rod was added to it.2g of the oil was weighed and poured into a glass-stopper bottle of about 250 ml capacity. 10ml of carbon tetrachloride was added to the oil to dissolve. 20ml of Wij's solution was added and a stopper was inserted and allowed to stay in the dark for 30minutes. 15ml of potassium iodide solution (10%) and 100ml of water was introduced and the mixture was thoroughly mixed and titrated with0.1M sodium thiosulphate solution using starch as indicator (titration = 'A'ml). A blank was carried out at the same time starting with 10ml of carbon tetrachloride (titration = 'B'ml)

$$IV = \frac{0.1269 \times (B-A) \times N \times 100}{W_{OIL}} \quad (4)$$

B= volume of sodium thiosulphate used in blank titration.
A= volume of sodium thiosulpate used in titration with oil.
N= normality of sodium thiosulphate.
W_{oil}= weight of oil used.

3.2.1.5. Determination of peroxide value

1g of oil was weighed into a clean drying boiling tube, 1g of powdered potassium iodide and 20ml of solvent mixture (2volume of glacial acetic acid + 1volume of chloroform) was added, the tube was placed in boiling water so that the liquid boils within 30seconds and was also allowed to boil vigorously for not more than 30seconds. The content was quickly poured into a flask containing 20ml of potassium iodide solution; the tube was washed out with 25ml of distilled water and was titrated with 0.002M sodium thiosulphate solution using starch as indicator. A blank was also carried out at the same time.

3.2.1.6. Determination of viscosity

A viscometer was used to determine the viscosity of the oil. Chloroform was poured first in the viscometer and the time at which the chloroform reached the bottom of the equipment was taken. The oil was then poured into the viscometer and the time taken far the groundnut oil to reach the bottom of the equipment was taken and recorded in mm^2/s

3.2.1.7. Determination of moisture content

In other to determine the moisture content in the oil (%), 48.15g of oil was weighed in a moisture pan, the weight of the pan and oil was taken and was put inside an oven for 3hours at a temperature of 45^{0}C. After every 1hour, the sample was cooled and weighed until the weight before and after was approximately equal.

3.2.1.8. Determination of refractive index

Abbe's refractometer was used to determine the moisture content of the oil sample, the equipment was well cleaned with cotton wool and placed in a bright room for light. The refractometer was calibrated using distilled water then small sample of the oil was placed on it and the refractive index observed was taken and recorded.

3.2.2. Two-step transesterification of the crude oils

In order to avoid the problem of saponification, a two-step method was used for synthesis of biodiesel from Jatropha curcas and Ricinus communis.

3.2.2.1. Acid catalyzed esterification

This is considered as a pre-treatment for the crude oils in order to reduce their water content which is the main cause of soap formation and subsequently, reduce its FFA. The oil was heated in the reaction glass tube to 65°C and a solution of concentrated H_2SO_4 acid (1.0% based on the oil weight) in methanol (30% v/v) was heated to 45°C and added to the reaction glass tubes. The resulting mixture was stirred on a magnetic stirrer for 1hr and the content was poured into a separating funnel and allowed to settle for 2 hours. The methanol-water fraction at the top layer was removed and the oil was decanted to be used for transesterification reaction.

3.2.2.2. Alkali catalyzed transesterification

50ml of oil was measured and poured into a 150ml conical flask and heated to a temperature of 45°C using a water bath. A solution of sodium methoxide was prepared in a 250ml beaker using 0.25g of NaOH pellet and 10.5mls of anhydrous methanol. The solution was properly stirred until the NaOH pellet was completely dissolved in it. The sodium methoxide solution was then poured into the warm oil and stirred vigorously for 90minutes using a magnetic stirrer and the mixture was left to settle for 24hours in a separating funnel. After settling, the upper layer which was biodiesel was decanted into a separate beaker while the lower layer which comprises of glycerol and soap was collected from the bottom of the funnel. The quantity of biodiesel collected was measured and recorded.

3.2.3. Washing

Biodiesel must be washed to remove any remaining methanol, glycerin, catalyst, soaps and other impurities. Water used is warmed to about 45°C and is passed through the esters to allow soluble material, excess catalyst and other impurities to stick to the water and be settled to the bottom of the vessel. The water is removed from the vessel periodically until the wash water drained out is clear or the pH of the biodiesel becomes relatively neutral.

3.2.4. Drying

The biodiesel washing sometimes leaves the biodiesel looking a bit cloudy. This means there's still a little water in it. It was heated slowly to 100°C and held there until all moisture present was evaporated i.e. dry.

3.2.5. Characterization of the biodiesel produced

3.2.5.1. Determination of specific gravity/density (ASTM D1298) by hydrometer method

This procedure is used to measure of specific gravity of the biodiesels. A clean dry empty 50ml density bottle is to be weighed and the mass recorded as M, it is then filled up with distilled water and subsequently with the samples. The mass of the bottle and water is taken and recorded as M_1 and that of biodiesel as M_2 respectively hence, the specific gravity is evaluated. This procedure is used to determine the specific gravity of the sample.

3.2.5.2. Determination of flash point: ASTM D 93 — Flash-point by Pensky-Martens closed cup tester

A sample of the biodiesel is heated in a close vessel and ignited. When the sample burns, the temperature is recorded; the pensky-martens cup tester measures the lowest temperature at which application of the test flame causes the vapor above the sample to ignite. The biodiesel is placed in a cup in such quantity as to just touch the prescribed mark on the interior of the cup. The cover is then fitted onto the position on the cup and Bunsen burner is used to supply heat to the apparatus at a rate of about 5°C per minute. During heating, the oil is constantly stirred. As the oil approaches its flashing, the injector burner is lighted and injected into the oil container after every 12 second intervals until a distinct flash is observed within the container. The temperature at which the flash occurred is then recorded, it is repeated three times and the average taken.

3.2.5.3. Determination of cloud point (ASTM D 2500)

A sample of the biodiesel is placed in a test jar to a mark and then placed inside a cooling bath. The temperature at the bottom of the test jar that is the temperature at which the biodiesel starts to form cloud is taken as the cloud point.

3.2.5.4. Determination of kinematic viscosity (ASTM D 445)

A viscometer is inserted into a water bath with a set temperature and left for 30minutes. The sample is added to the viscometer and allowed to remain in the bath as long as it reaches the test thermometer. The sample is allowed to flow freely and the time required for the meniscus to pass from the first to the second timing mark is taken using a stop watch. The procedure is repeated a number of times and the average value are taken which is then multiplied with the viscometer calibration to give the kinematic viscosity.

3.2.5.5. Determination of pour point (ASTM D 97)

A sample of the biodiesel is kept in the freezer to about 50^{0}C then placed in a heating mantle to melt. The temperature at the bottom of the test jar that is the temperature at which the biodiesel starts to pour is taken as the pour point.

3.2.5.6. Cetane Number (ASTM D 613)

Cetane Number is a measure of the fuel's ignition delay. Higher Cetane numbers indicate shorter times between the injection of the fuel and its ignition. Higher numbers have been associated with reduced engine roughness and with lower starting temperatures for engines.

3.2.5.7. Acid Number (ASTM D 664)

The biodiesel sample is measured and poured in a beaker. A neutral solvent (a mixture of petroleum ether and ethanol) is prepared and 50ml of it is taken and poured into the beaker containing the biodiesel. The mixture is stirred vigorously for 30minutes. 0.56g of potassium hydroxide (KOH) pellet is measured and placed in a separate beaker and 0.1M KOH is prepared, 3drops of phenolphthalein indicator is added to the sample and is titrated against 0.1M KOH till the color change observed turned pink.

3.2.5.8. Determination of sulphur content (ASTM D 5453)

The sulphur content was determined by the energy dispersive X-ray fluorescence spectroscopy technique. The biodiesel was placed in disposable plate covered with male and female cells; the sample was placed in a oil in sulphur test equipment and left for 10minutes. The equipment measures the sulphur content of the biodiesel three (3) consecutive times and takes the average which is then recorded as the sulphur content.

3.2.5.9. Determination of distillation characteristics (ASTM D 86)

The distillation characteristics were studied using distillation apparatus (Model PMD 100) in accordance to the procedure in **ASTM D 86**. 100ml of the biodiesel was charged into the distillation flask. A thermometer provided with a snug-fitting cork was tightly fitted into the neck of the distillation flask. The flask was then fixed tightly into the condenser tube by raising and adjusting the flask support board of a calibrated distillation batch unit. Systematic observations and recordings of temperature readings at 5mls, 10mls, 20mls, 30mls, 40mls, 50mls, 60mls, 70mls, 80mls, and 90mls respectively, and volumes of condensate were taken and recorded.

3.2.5.10. Determination of bottom water and sediment (ASTM D 2709)

The water and sediment test was determined according to **ASTM D 2709**, 50mls of the biodiesel and 50ml of toluene were mixed in a 100ml centrifuge tube with the tube tip having graduation of 0.01ml over the range of 0 to 0.2ml, the centrifuge tube was shaken so that an even distribution of the mixture is observed. The tube was placed in a trunnion cup inside an oil test centrifuge in such a way that the tubes are placed opposite each other to establish a balance in the centrifuge. The centrifuge is then closed and timed for 30 minutes; the samples are then whirled for agitation at a speed of 1800rpm to ensure homogeneity. The combined water and sediment at the bottom of the tube was reported to the nearest 0.005ml.

4. Results and discussion of results

Concern on the situation of fossil fuel as the sole sources of energy and environmental pollution emanated from the utilization of fossil fuel resulted into considerable attention given to alternative sources of energy. Biodiesel is considered as a perfect alternative source of energy that can compete with fossil fuel in terms of performance and efficiency. However, production of biodiesel from vegetable oils and animal fats also raised a serious concern of food crisis and the critics of the production of biodiesel from edible are making it difficult to justify the use of edible oil for fuel, considering the tremendously increment in demand for edible oil. Hence the need to produce biodiesel from non-edible oils such as Jatropha Curcas and Ricinus Communis oil that are grown in large quantities and waste lands in Nigeria, which is the focus of this study. Results obtained on characterization of the oils and biodiesel produced from the non-edible oils are hereby presented.

4.1. Physiochemical properties of Jatropha curcas and Ricinus communis oil

Ricinus communis and Jatropha Curcas oils that were utilised as a feedstock in the production of biodiesel were characterized to determine their physiochemical properties such as density, pH, viscosity, acid value, iodine value etc. The results obtained on the properties of oils were compared with that of American Oil Chemists' Society (AOCS) standard values and the results obtained are presented in Table 1. The iodine value shows the level of unsaturation of the oil and also influences the oxidation and deposition formed in diesel engines. It is used in determining the drying property of the oil. Iodine value obtained for both Jatropha curcas and Ricinus communis were seen to be $98gI_2/100g$ and $85gI_2/100g$ respectively and both values fall within the acceptable AOCS limit as shown in Table 1. The iodine value of Jatropha ($98\ gI_2/100g$) was seen to be higher than that of Ricinus ($85\ gI_2/100g$) and this signifies that there is a higher degree of unstauration in the former than in the latter, although, both oils are classified as non-drying oils since their iodine value is below $115\ gI_2/100g$. Moisture in vegetable oils is a great impediment to the formation of esters due to increase in tendency of soap formation and thus will have to be minimal for transesterification to occur [14]. The moisture content in both Jatropha curcas and Ricinus communis were seen to be 0.20% and 0.24% respectively and are within the range specified by AOCS as shown in Table 4.0. The moisture content of Jatropha curcas (0.20%) was seen to be a little less than that of Ricinus communis (0.24%) and this signifies a higher content of water in Ricinus communis than Jatropha curcas. The saponification value of oil is a measure of the tendency of the oil to form soap during the transesterification reaction. The saponification value obtained for Jatropha curcas and Ricinus communis were seen to be 190mgKOH/g and 178mgKOH/g respectively and are within the range specified by AOCS as shown in Table 1. The saponification value of Jatropha curcas (190mgKOH/g) was seen to be a bit higher than that of Ricinus communis and this thus signifies a higher tendency of the former forming soap during transesterification reaction than the latter. Vegetable oils with high acid value are classified as inedible while those with low acid value are classified as edible oils. The acid value obtained for both Jatropha curcas and Ricinus communis were seen to be 36.8mgKOH/g and 0.913mgKOH/g respectively and both values fall within the acceptable AOCS limit as shown in Table 1. Leung and Guo (2006) [56] reported that oils with high acid value tend to deactivate catalyst used during transesterification and the acid value of Jatropha (36.8mgKOH/g) was seen to be exceedingly higher than that of Ricinus

(0.913mgKOH/g) and this signifies that the ester yield of Jatropha curcas will be less than that of Ricinus communis due to increase catalyst deactivation and soap formation in the former. Also, the same can be said for the free fatty acid (FFA) present in both with their values being 18.4% and 0.457% respectively. Kinematic viscosity which is a measure of the flow capabilities of the oils was measured at 30°C. The kinematic viscosity of Jatropha curcas and Ricinus communis were seen to be 51.8mm²/s and 6.9mm²/s respectively and both are within the range specified by AOCS as shown in Table 1. The kinematic viscosity of Jatropha curcas (51.8mm²/s) was seen to be extremely higher than that of Ricinus communis (6.9mm²/s) and this means that Ricinus communis has a higher flow capability than Jatropha curcas and can easily undergo transesterification.

Parameters	Unit	Obtained Expt Value for Jatropha Curcas	Obtained Expt Value for Ricinus Communis	AOCS Standard Value for Jatropha Curcas	AOCS Standard Value for Ricinus Communis
Saponification Value (S.V)	mgKOH/g	190	178	188-195	176-184
Iodine Value (I.V)	gI_2/100g oil	98	85	84-100	83-88
Acid Value (A.V)	mgKOH/g	36.8	0.913	1.0-38.2	2.0 max
Free Fatty Acid (FFA)	%	18.4	0.457	<1%	<1%
Specific Gravity (S.G) at 30 °C	-	0.913	0.89	0.910-0.915	0.88-0.915
Density (ρ)	g/ml		0.856	-	-
Refractive Index (R.I) at 30 °C	-	1.466	1.468	1.467-1.470	1.467-1470
Kinematic Viscosity (v) at 30 °C	mm²/s	51.8	6.9	-	6.3-8.9
Moisture Content	%	0.20	0.24	-	0.355 % max

Table 1. Comparison of Physiochemical Properties of Ricinus communis and Jatropha curcas oils with AOCS standard values

4.2. Optimization of two-step transesterification of Jatropha curcas and Ricinus communis

Transesterification is the reaction between triglycerides and lower alcohols to produce free glycerols and the fatty acid ester. It involves reaction between the oil which is the feedstock and an alcohol, usually methanol in the presence of a catalyst such as sodium or potassium hydroxide to give corresponding esters. Although, the main factors influences the transesterification reactions are the alcohol to oil molar ratio, catalyst type and concentration, reaction temperature and reaction time, with the methanol to oil ratio higher than stoichiometrix ratio to drive the equilibrium to a maximum ester yields. However, in this present study all other factors are kept constant except the temperatures that are varying between 50°C to 60°C. Sodium hydroxide was selected as the catalyst, the ratio of oil to methanol was 1:6, reaction time was 60 minutes and the weight of catalyst was 0.5wt%. Results obtained on the effect of temperature on the yield of biodiesel from Jatropha Carcus and Ricinus Communis oil as the feed stock are presented in Table 2. The best yields were obtained for both the methyl ester of Jatropha curcas and Ricinus communis at the temperature of 60°C with the percentage methyl ester yield being 96% and 98%. This is because at 60°C, the molecules of the triglycerides of Jatropha curcas and Ricinus communis had high kinetic energy and this thus increased the collision rate and therefore, improved the overall process by favouring the formation of methyl esters while at the lower temperatures of 50°C and 55°C with corresponding percentage methyl ester yield of 86% and 90% and 92% and 95% respectively, there was lesser collision of reacting molecules and thus, reduced biodiesel yield as seen in Table 2. The reason for this behaviour is due to the endothermic nature of the reaction. The higher reaction temperature would favour endothermic reaction, thus increasing the rate of reaction as well as the ester concentration. Although at 65°C, it was noticed that there was a drop in percentage methyl ester yield for both Jatropha curcas and Ricinus communis as shown in Fig 4.0 below with their corresponding values being 84% and 86% respectively. This is because at the reaction temperature (65°C), there was increased vaporization of the alcohol (methanol) used in the transesterification process due to the proximity of the reaction temperature (65°C) to the boiling point of methanol (64.7°C). Based on the results obtained on the effect of temperature on the yield of biodiesel, it can be deduced that the optimum conditions for the production of methyl ester from crude Jatropha curcas and Ricinus communis through two-stage transesterification are oil to methanol molar ratio of 1:6, catalyst concentration of 0.5wt%, reaction time of 60 minutes and reaction temperature of 60°C.

Temperature (°C)	Jatropha curcas Production Yield (wt%)	Jatropha curcas Methyl Ester Yield (wt%)	Ricinus communis Production Yield (wt%)	Ricinus communis Methyl Ester Yield (wt%)
50	89	86	93	90
55	94	92	98	95
60	99	96	99	98
65	90	84	92	86

Table 2. Effect of Temperature on Biodiesel Yield from Jatropha curcas and Ricinus communis

4.3. Characterization of produced esters

The biodiesel produced from Jatropha curcas and Ricinus communis using the optimal conditions of 1:6 oil to methanol molar ratio, 0.5wt% catalyst concentration, 60 minutes reaction time and reaction temperature of 60°C were analyzed for their fuel properties. Their properties were compared to the ASTM D 6751 biodiesel standard and the ASTM D 975 fossil diesel standard in order to confirm their acceptability as fuel in diesel engines as shown in Table 3. Kinematic viscosity is a measure of resistance to flow of a liquid due to internal friction of one part of the fluid moving over another. High viscosity affects the atomization of a fuel upon injection into the combustion chamber and thus leads to the formation of engine deposits [57]. The kinematic viscosities of Jatropha curcas and Ricinus communis were determined to be 4.93mm^2/s and 14.2mm^2/s respectively and when compared to ASTM D 6571, it was seen that the viscosity of Jatropha falls within range while that of Ricinus communis is far from the acceptable standard, although, it falls within the acceptable limit of fossil diesel (ASTM D 975) as presented in Table 3. It can be therefore said that the methyl esters of Ricinus communis has a higher tendency of forming deposits in engines than that of Jatropha curcas due to the significant difference in kinematic viscosity and this makes Jatropha curcas methyl ester a more suitable fuel as regards viscous property. The flash point is the lowest temperature at which an applied ignition source will cause the vapours of the fuel to ignite. It is therefore a measure of tendency of a sample to form a flammable mixture with air. The flash point obtained for Jatropha curcas and Ricinus communis were 142°F and 150°F respectively and they fall within the range of standard limit set by ASTM D 6751 (130°F min) [58]. Ricinus communis has a higher flash point when compared to Jatropha curcas, but both are considerably high and this leads to their safer handling and storage. Cloud point and pour point have implications on the use of biodiesel in cold weather applications. The cloud point is the most common measure of the tendency of a fuel to crystallize and the cloud point of Jatropha curcas and Ricinus communis were determined as 5°C and 8°C respectively. This signifies that Ricinus communis has a high tendency of forming cloudy crystals easily in cold temperature than Jatropha curcas.

Biodiesel is also potentially subject to hydrolytic degradation caused by the presence of water. Fuel contaminated with water can cause engine corrosion and breakdown. From the results obtained, Jatropha has traces of bottom sediment and water while Ricinus communis has 0.05%vol and both values conforms to the ASTM D 6751 set standard whose maximum allowable limit is 0.05%vol. Since Jatropha curcas methyl ester has a neglible amount of bottom water and sediment, it is of better quality compared to Ricinus communis. Sulfur in the atmosphere has negative impacts on human health and on the environment and biodiesel have traditionally been acknowledged as sulfur-free which is of great advantage over fossil diesel and the results obtained confirms that fact. The sulfur content of both Jatropha curcas and Ricinus communis were 0.05%wt and 0.03%wt which is within the ASTM D 6751 standard limit (0.05 max). Also, the sulphur content of the methyl ester of Jatropha was seen to be higher than that of Ricinus and this signifies that the latter is more environmentally friendly than the former. Elevated total glycerine values are indicators of incomplete esterification reactions and predictors of excessive carbon deposits in the engine. This can be as a result of incomplete washing of the ester after production. The free glycerine is a source of carbon deposits in the engine because of incomplete combustion. The free and total glycerines determined for the methyl ester of Jatropha curcas and Ricinus communis were 0.02wt% and 0.24wt% and 0.02wt% and 0.23wt% respectively and both values were seen to be within the acceptable standard of ASTM D 6751. On comparison with each other, they were found to be approximately equal. Generally, the cetane number is a dimensionless descriptor of the ignition quality of a diesel fuel. As such, it is a prime indicator of diesel fuel quality. The cetane number of the methyl esters of Jatropha curcas and Ricinus communis were found to be 46.11 and 50.64 respectively. On comparison with standard, it was found that the cetane number of that of Jatropha curcas was below the allowable minimum of 47 stated by ASTM D 6751 while that of Ricinus communis was slightly above it. On further comparison with fossil diesel standard, it was seen that the centane number of both was above the allowable minimum of 40 imposed by ASTM D 975. Thus results confirm the fact that biodiesels generally, have a higher cetane number than fossil diesels

Property	Unit	ASTM Test Method	Expt. Value for Jatropha curcas	Expt. Value for Ricinus communis	ASTM Standard for Biodiesel (ASTM D 6751)	ASTM Standard for Petrol diesel (ASTM D 975)
Kinematic Viscosity at 40°C	mm^2/s	D 445	4.39	5.42	1.9-6.0	1.9-4.1
Density/Specific Gravity	kg/l	D 1298	0.88	0.88	0.86-0.89	0.95 max
Flash Point	°F	D 93	142.0	150.0	93.0 min	150 min
Cloud Point	°F	D 2600	41.0	46.4		40 max
Total Sulphur (X-Ray)	% wt	D 4294	0.05	0.03	0.05max	0.50 max
Bottom Water & Sediment	% vol	D 1796	Trace	0.05	0.05 max	0.50 max
Distillation Properties IBP	°C	D 86	128	302	360 max	-
FBP	°C		329	360		205 max
Total Recovery	%		98	97		
Centane Number	-	D 975	46.11	50.64	47 min	40 min
Water by Distillation	% vol	D 95	Trace	0.05	-	0.5 max
Free Glycerine	% mass	D 6584	0.02	0.02	0.02 max	0.02 max
Total Glycerine	% mass	D 6584	0.24	0.23	0.24 max	0.24 max

Table 3. Fuel Properties of Jatropha curcas and Ricinus communis Methyl Esters

Boiling point is the temperature at which a liquid transition to a gas and it is also related to the flash point. For a pure substance the boiling point is a single temperature value.

However, for a mixture of hydrocarbons as existing in biodiesel fuel, there is a range of boiling points for the different constituent chemical specie. The data tested for in the methyl ester of Jatropha curcas and Ricinus communis includes the initial boiling point (IBP), boiling temperature corresponding to increments of the volume of fuel distilled (10%, 20%, 30%, 40%, 50%, 60%, 70%, 80% and 90%), final boiling point (FBP) and % total recovery (Table 4) .

Percentage Recovery (cm^3)	Temperature (°C)	
	Jatropha curcas	Ricinus communis
IBP	128	302
10%	234	307
20%	255	309
30%	260	312
40%	290	322
50%	313	330
60%	321	336
70%	326	343
80%	328	350
90%	329	360
FBP	329	362
% Total Recovery	97.77	97.00

Table 4. Distillation Characteristic Table of Jatropha curcas and Ricinus communis Methyl Esters

5. Conclusion

The need to move away from oil as a major source of energy is growing every year, due to the price instability and environmental pollution which are the consequence of over dependence on the fossil fuel. The current research and development on alternative energy that will replace fossil fuel is now focusing on the biofuel. Renewable biofuel will reduce the dependence on oil and also reduce the trade deficit of nations, especially the developing nations. However, for biofuel to compete with the fossil fuel and possibility replacing it as the energy source, needs to be easy, cheap and fast to produce. Production of biofuel from vegetable oil is however not economical due to the fact that production of biodiesel from vegetable will lead to food crisis while trying to solve energy crisis. It is therefore much more desirable to use non-edible oils as a feedstock in the production of biodiesel, which is the focus of this study. This study report the production of biodiesel from non-edible oils (Jatropha Carcus and Ricinus Communis) as alternative to petrol diesel. Based on the results of experimental analysis, it can be concluded that oils from both feed stocks are suitable for the production of biodiesel. Also, temperature had a high effect on biodiesel yield and the yield increased with increasing temperature up to a point where the reacting temperature was proximal to the boiling point of the methanol. It can also be concluded that, though, Ricinus communis had a higher methyl ester yield (98wt%) than Jatropha Carcus (96wt%) at same optimal reaction conditions, the methyl ester of Jatropha Carcus is a more suitable fuel for diesel engines than that of Ricinus communis because the fuel properties obtained for the former were more compatible with the engines and in accordance with ASTM D 6751 than that of the latter.

Author details

A.S. Abdulkareem*

Department of Chemical Engineering, School of Engineering and Engineering Technology, Federal University of Technology, PMB 65 Minna, Niger State, Nigeria

Department of Civil and Chemical Engineering, College of Science, Engineering and Technology, University of South Africa, Private Bag X6, Florida 1710, Johannesburg, South Africa

A. Jimoh, J.O. Odigure and D. Patience

Department of Chemical Engineering, School of Engineering and Engineering Technology, Federal University of Technology, PMB 65 Minna, Niger State, Nigeria

A.S. Afolabi

Department of Civil and Chemical Engineering, College of Science, Engineering and Technology, University of South Africa, Private Bag X6, Florida 1710, Johannesburg. South Africa

Acknowledgement

Support received from Step B Project, Federal University of Technology, Minna Nigeria is highly appreciated. National research foundation (NRF), South Africa and Faculty of Science, Engineering and Technology, University of South Africa are also appreciated for their support.

6. References

[1] Abdulkareem, A.S.; Odigure, J.O. & Kuranga. M.B. (2010). Production and Characterization of Bio-Fuel from Coconut oil. Energy Source Part A. J. 32 106-114.

[2] Adeniyi, O.D.; Kovo, A.S.; Abdulkareem. A.S & Chukwudozie. C. (2007): Ethanol Fuel Production from Cassava as a Substitute for Gasoline. Dispersion and Technology.J. 28 501-504.

[3] Carraretto, C.; Macor, A. & Mirandola, A. (2004). Biodiesel as alternative fuel: Experimental analysis and energetic evaluation. Energy. J. 2195-2211.

[4] Gerpen, V.J. (2005). Biodiesel Processing and Production. Fuel Processing Technology. J. 86(10) 1097–1107.

[5] Abdulakreem, A. S & Odigure, J.O. (2002). Radiative Heat Evaluation from Gas Flaring By Computer Simulation. Association for the advancement of Modelling and simulation in enterprises, Lyon France. J. 71 No 2 19 – 35.

* Corresponding Author

[6] Abdulkareem, A. S. (2005). Evaluation of ground level concentration of pollutant due to gas flaring by computer simulation: A case study of Niger – Delta area of Nigeria. Leonardo Electronic Journal of Practices and Technologies, Technical University of Cluj - Napoca Romania. J. 6 29 – 42.

[7] Abdulkareem, A. S.(2005). Urban Air Pollution Evaluation by Computer Simulation: A Case study of Petroleum Refining Company, Nigeria. Leonardo Journal of Science Technical University of Cluj - Napoca Romania. J. 6 17 – 28.

[8] Abdulkareem, A.S & Odigure, J.O. (2006). Deterministic Model for Noise Dispersion from gas Flaring: A case study of Niger – Delta area of Nigeria. Chemical and Biochemical Engineering, Croatia. J. Q 20, No 2 139 – 146.

[9] Agarwal. A.K & Das. L.M. (2001). Biodiesel development and characterization for use as a fuel in compression ignition engines. Engineering Gas Turbines Power.J. 123 440-447.

[10] Abdulkareem, A.S.; Idibie, C.A.; Afolabi, A.S.; Pienaar, H.C.vZ. & Iyuke S.E. (2010): Kinetics of sulphonation of polystyrene butadiene rubber in sulphuric acid medium. International Review of Chemical Engineering. J. 2, No7 832-839.

[11] Ayhan. D. (2008). Importance of biomass energy sources for Turkey. Energy policy. J. 36 834-842.

[12] Barminas J. T., Maina H. M., Tahir S., Kubmarawa D., Tsware K., (2001). A preliminary investigation into the biofuel characteristics of tigernut (cyperus esculentus), Bioresources. Technology. J. 79 87-89.

[13] U.S. Department of Energy, Energy Efficiency and Renewable Energy, "Biodiesel Handling and Use Guidelines", October 2004.

[14] Abdulkareem A.S., Uthman H., Afolabi A.S and Awonebe O.L (2011). Extraction and Optimization of Oil from Moringa Oleifera Seed as an Alternative Feedstock for the production of Biodeisel. Majid N, Mostafa K, editors. Sustainable Growth and Application in Renewable Energy Sources. InTech. Pp243-268.

[15] Ahmmad, M.; Ullah, K.; Khan, M.A.; Zafari, M.; Tariq, M.; Ali, S. & Sultana. S. (2011). Physico chemical analysis of hemp biodiesel: A Promising non edible new sources for bioenergy. Energy Sources, Part A. J. 33 1365-1374.

[16] Aghan D. (2005). Biodiesel production from vegetable oils via catalytic and non catalytic supercritical methanol tranestarification methods. rogress in energy and combustion. J. 31 406-487.

[17] Dalai A.K & Kulkarni, M.G. (2006). Waste cooking oil-an economical source for biodiesel: A review. Industrial & Engineering Chemistry Research. J. 45 2901-2913.

[18] Eaves, J. and Eaves, S. (2007) Renewable corn-ethanol and energy security. Energy Policy. J. 5 5958–5963.

[19] Helwani, Z.; Othman, M. R,; Aziz, N.; Fernando, W. J. N.& Kim. J.(2009). Technologies for production of biodiesel focusing on green catalytic techniques: A review. Fuel Processing Technology.J. 90 1502-1514.

[20] El-Sabagh, S.M.; Keera, S.T. & Tama, A.R. (2011). The characterization of biodiesel fuel from waste frying oil. Energy Source, Part A .J. 33 401-409.

[21] Eevera, T.; Balamurughan, P.; and Chittibabu .S. (2011). Characterization of groudnut oil based biodiesel to assess the feasibility for power generation. Energy Sources, Part A. J. 33 1354-1364.

[22] Hamamci, C.; Saydut, A.; Tonbul, Y.; Kaya, C. & Kafadari, A.B (2011). Biodiesel production via transestarification from safflower (Carthemustinctorious L) seed oil. Energy Sources Part A. J. 33 512-520.

[23] Hossain, A.B & Boyce, A.N. (2009). Biodiesel production from waste sunflower cooking oil as an environmental recycling process and renewable energy. Bulgarian. Agricultural Science. J. 15(4) 312-317.

[24] Khunrong, T.; Punsuvon, A.; Vaithanomisate, P. & Pomchaitawand. C. (2011). Production of ethanol from from pulp obtained by steam explosion pretreatment of oil palm trunk. Energy Sources, Part A. J. 33 221-228.

[25] Clean cities energy, Report of U.S Department of Energy: Energy Efficiency and Renewable energy. 1-17.

[26] Ahmad, M.; Samuel, S.; Zafari, M.; Khan, M.A.; Tariq. M.; Ali. S. & Sultana. S. (2011). Physicochemical characterization of eco-friendly rice bran oil biodiesel. Energy, Part A. J. 33 1386-1397.

[27] Arjun,B. C.; Martin, S.T,; Suzanne,M. B.; Chris. W.& Rafiqul Islam. M. (2008).Non-Edible Plant Oils as New Sources for Biodiesel Production. Journal of Molecular Science. J. 9 No 2 169-180.

[28] Canakci, M. (2007): The potential of restaurant waste lipids as biodiesel feed stocks. Bioresource Technology. J. 98 183–190.

[29] Ahmmad, M.; Ullah, K.; Khan, M.A.; Ali, S.; Zafari, M. & Sultana. S. (2011). Quantitative and qualitative anaysis of sesame oil biodiesel. Energy Sources, Part A. J. 33 1239-1249.

[30] Durosoy, I.; Turker, M. F.; Keles, S. & Kaygusuz, K. (2011). Sustainable agriculture and the production of biomass for energy use. Energy Sources, Part A. J. 33 938-947.

[31] Refaat, A. A. (2010). Different techniques for the production of biodiesel from waste vegetable oil. International Journal of Environmental Technology. 7 (1) pp 183-213.

[32] Dokwadanyi, P (2011). Production and charcterization of biodiesel from Jatropha curcas and Ricinus communis. B.Eng project submited to the Department of Chemical Engineering, Federal Unuversity of Technology, Minna. Nigeria (Unpublished). 1-81.

[33] Hossain, M., Janaiah, A and Otsuka, K (2005). Is the productivity impact of the Green Revolution in rice vanishing? Economic and Political Weekly (31, December), 141-149.

[34] Shao, H. & Chu. L. (2008). Resources evaluation of typical energy plants and possible functional zone planning in China. Biomass and Bioenergy. J. 32 283-288.

[35] Smith K. and Edwards R. (2011) 2008 - The year of global food crisis. Herald Scotland, Sunday Herald.

[36] Peterson, C.L., Wagner, G.L. and Auld, D.L. (1983). Vegetable Oil Substitutes for Diesel Fuel, Transactions of the ASAE, 26(2): pp. 322- 327.

[37] Freedman B, Pryde EH, Mounts TL (1984). Variables affecting the yields of fatty esters from transesterified vegetable oils, Am. Oil Chem. Soc. J. 61 1638-1643.

[38] Tyson S (2001). Biodiesel handling and use guideline NREL Report, No TP-580-3004. Golden Co

[39] Schumacher, L.G., Clark N, N., Lyons D.W and Marshall W (2011). Diesel Engine Exhaust Emissions Evaluation of Biodiesel Blends, Using a Cummins 110e Engine. Trans ASAE. J. 44 (6) 1461-1464.

[40] Bowman, M.; Hilligoss, D.; Rasmussen, S. and Thomas. R (2006).Biodiesel: A Renewable and Biodegradable Fuel. Hydrocarbon Processing. J. 103-106.

[41] Pasqualino, J.C.; Montane, D. & Salvado, J (2006). Synergic effects of biodiesel in the egradability of fossil-derived fuels. Biomass & Bioenergy. J. 30 pp. 874–879.

[42] Werner K (2002): New trends in developing biodiesel world-wide. Asia Bio-Fuels: Evaluating the Commercial Uses of Ethanol. Conference of Alcohol, & Biodiesel. Singapore, 22-23 April, 2002.

[43] Graboski, M.S., McCormick, R.L. (1998). Combustion of fat and vegetable oil derived fuels in diesel engines. Progress in Energy Combustion Science. J. 24, 125–164.

[44] Korbitz, W. (1999). Biodiesel production in Europe and North America, an encouraging prospect. Renewable Energy. J. 16(1-4), 1078-1083.

[45] Worgetter, M.; Prankl, H. & Rathbauer, J. Eigenschaften von Biodiesel (1998). Landbauforschung Völkenrode. Sonderheft 190 (Biodiesel- Optimierungspotentiale und Umwelteffekte) 31-43.

[46] Patzek, T. W. (2004). Thermodynamics of the corn-ethanol biofuel cycle, Critical Reviews in Plant Sciences 23(6): 519–567, An updated web version is at http://-petroleum.berkeley.edu/papers/patzek/CRPS416-Patzek-Web.pdf

[47] Boopathy, R., (2004). Anaerobic biodegradation of no. 2 diesel fuel in soil: A soil column study. Bioresour Technol. J. 94:143–51..

[48] Claude, S(1999). Research of new outlets for glycerol-recent developments in France. Fett/Lipid. J. 101 (3) 101–104.

[49] Arndt, C. and K.R. Simler (2007), 'Consistent Poverty Comparisons and Inference', Agricultural Economics. J. 37 133-143.

[50] Refaat, A.A.; El Sheltawy, S.T. & Sadek, K.U. (2008). Optimum reaction time, performance and exhaust emissions of biodiesel Produced by microwave irradiation. International Journal of Environmental Science and Technology. J. 5 No3 315-322.

[51] Sarin, R.; Sharma, M.; Sinharay, S.; Malhtra, R. K., (2006). Jatropha – Palm biodiesel blends an optimum mix for Asia. Fuel 86 (10-11) 1365-1371.

[52] Azan M. M., Waris A and Nahar N.M (2005). Prospects and potential of fatty acid esters of some non- traditional seed oils for use as biodiesel in India. Biomass and Bioenerg. J. 24, 293-302

[53] Tiwari, A. K.; Kumar, A.; Raheman, H., (2007). Biodiesel production from jatropha oil (*Jatropha curcas*) with high free fatty acid: An optimized process. Biomass Bioenerg. J. 31 (8) 569 -575.

[54] Berchamns H.J and Hirata S (2008). Biodiesel Production from Crude Jatropha Curcas L Seed Oil with a high content of free fatty acids. Bioresources Technology. J. 99 1716-1721.

[55] Halder, P., Havu-Nuutinen, S., Pietarinen, J. & Pelkonen, P.(2011). Bio-energy and youth: Analyzing the role of school, home, and media from the future policy

perspectives. Applied Energy 88 (2011) 1233–1240. Retrieved from http://www.sciencedirect.com.ludwig.lub.lu.se/science?.

[56] Leung DYC, Guo Y. 2006. Transesterification of neat and used frying oil: optimization for biodiesel production. Fuel Process Technol. J. 87 883–890.

[57] Allen, C. W., K. C. Watts, and R. G. Ackman. 1999. Predicting the surface tension of biodiesel fuels from their fatty acidcomposition. American Oil Chem. Soc. J. 76(3) 317-323.

[58] ASTM D-6751-03a (2003). Standard Specification for Biodiesel Fuel Blend Stock (B100) for Middle Distillate Fuels. ASTM International. 1-6.

Chapter 8

Production and Characterization of Biofuel from Refined Groundnut Oil

A. Jimoh, A.S. Abdulkareem, A.S. Afolabi, J.O. Odigure and U.C. Odili

Additional information is available at the end of the chapter

http://dx.doi.org/10.5772/52443

1. Introduction

Persistent global energy and fuel crises have shown that fossil fuels are limited source of energy that will ultimately get exhausted [1-3]. Continuous depreciation of the world oil reserves as reported by USEIA (2007) [4] also corroborates the fact that fossil fuels resources like petroleum, coal and natural gas are finite and non-renewable source of energy. Apart from the foreseen shortage of fossil fuels, its rising price and increasing difficulty of paying for them in the next years to come, necessitate the serious urge for alternative resource that will offers reduced emissions, improved biodegradability, improve performance and clean up emissions necessitated continuous research into the development of renewable energy source. Aside the fact that gasoline and diesel are the most widely used fuels coming from fossil fuel [5], it is a known fact now that gasoline contributes to increase hazardous emissions [6] and that diesel with higher carbon numbers contributes to emissions of high particulate matters, high sulphur dioxide and high poly aromatic hydrocarbons [5]. Thus, need for the kind of serious attention being witness in the search for economically viable and environmental friendly renewable fuels like biodiesel.

Biodiesel has been reported to have offered reduced exhaust emissions, improved biodegradability, reduced toxicity and higher cetane rating which can improve performance and clean up emissions [7]. More so, vegetable oil has good resource for the production of biodiesel through transesterification of different types like canola, rapeseed, soybean oil, rapeseed etc [8]. Interesting results have been reported on the production of biodiesel through transesterification of different kinds of vegetable oil from different parts of the world, such include soybean (United State), rapeseed (Europe), oil palm (South-East Asia), jatropha curcus and rice bran oil (India) [9-13]. Besides, esters from vegetable oils have been reported as the best substitutes for diesel because they do not demand any modification in the diesel engine and have a high energetic yield [14]. Fossil fuels dependent countries like

Nigeria have commenced initiation of policies aimed at re-directing the nation's major energy sources from the finite crude oil to renewable sources. Interestingly, Nigeria is endowed with different renewable energy resources to support the implementation of such policies as reported by [5, 15]. Literature have shown that vegetable oils are one of the significant renewable feedstock that can provides secure, abundant, cost effective and clean source of energy for developing nations like Nigeria. Common vegetable oil in the country include palm oil, palm kernel oil (PKO), groundnut oil, cottonseed and soybean [5]. Considerable works have been reported in print on biodiesel production from vegetable oils but limited studies were found for vegetable oils common in Nigeria. As at year 2000, limited studies were reported in the literature on production and testing of biodiesel from Nigerian Lauric oils [5]. Recent development only shows some improvement on the exploitation of Nigeria vegetable oils as resources for biodiesel production, such include transesterification of palm kernel oil [5, 16], transesterification of groundnut oil [17-19]. With increasing attention on the use of these oils for biodiesel production, the present study focused more on the optimization of biodiesel production from refined groundnut oil relative to the effects of the major reaction parameters. Groundnut oil has been reported to be among the most consumed vegetable oil and the 13th most important food crop in the world. It is also known to contain a high percentage of unsaturated fatty acid namely oleic and linoleic acid [20] and low free fatty acid value [21].Groundnut seed contain high quality edible oils, protein, and carbohydrate. Globally, 50% of groundnut produced is used for oil extraction, 37% for confectionary use and 12% for seed purpose. The fat in the oil is approximately 50% monounsaturated, and 30% polyunsaturated [22]. In comparison, groundnut oil has been reported of having capacity to produce approximately 123 gallons of biodiesel per acre while soybeans yield only 50 gallons [23]. This research work is therefore aimed at optimizing the production of biodiesel from groundnut oil due to its cheapness and availability. This aim can be achieved through the realization of the following objectives;

i. Characterization of the commercially obtained refined groundnut oil by testing for properties like Specific gravity (or density), flash point, kinematic viscosity, acid value, iodine value, sulphur content, moisture or water content etc. Comparison of the results with literature values.
ii. Conversion of the groundnut oil into biodiesel through alkali-based transesterification with methanol using sodium hydroxide (NaOH) as catalyst.
iii. Detailed characterization of the properties of biodiesel produced and comparison of the experimental results obtained with available standards.
iv. Investigation of the effects of temperature, reaction time, catalyst concentration, and mass ratio of methanol to groundnut oil on production yield of the biodiesel.

1.1. First biodisel

Walton J., in 1938 recommended that "to get the utmost value from vegetable oils as fuel, it is academically necessary to split off the triglycerides and to run on the residual fatty acid. Practical experiments have not yet been carried out with this; the problems are likely to be much more difficult when using free fatty acids than when using the oils straight from the crushing mill. It is obvious that the glycerides have no fuel value and in addition are likely to cause an excess of carbon in comparison with gas oil. Walton's statement pointed in the direction of what is now termed "biodiesel" by recommending the elimination of glycerol from the fuel, although esters was not mentioned. In this connection, some remarkable work performed in Belgium and its former colony, the Belgian Congo (known after its independence for a long time as Zaire), deserves more recognition than it has received. It appears that Belgian patent 422,877, granted on Aug. 31, 1937 to Chavanne G. (University of Brussels, Belgium), constitutes the first report on what is today known as biodiesel. It describes the use of ethyl esters of palm oil (although other oils and methyl esters was mentioned) as diesel fuel. These esters were obtained by acid-catalyzed transesterification of the oil (base catalysis is now more common). This work was of particular interest as it was related to an extensive report published in 1942 on the production and use of palm oil ethyl ester as fuel and the work described the first test of an urban bus operating on biodiesel. A bus fueled with palm oil ethyl ester served the commercial passenger line between Brussels and Louvain (Leuven) in the summer of 1938. The performance of the bus operating on that fuel was reportedly satisfactory. It was noted that the viscosity difference between the esters and conventional diesel fuel was considerably less than that between the parent oil and conventional diesel fuel [24]. The patent also pointed out that the esters are miscible with other fuels and the work also discussed the cetane number (CN) testing of a biodiesel fuel. In the report, the CN of palm oil ethyl ester was reported as ~83 (relative to a high-quality standard with CN 70.5, a low-quality standard of CN 18, and diesel fuels with CN of 50 and 57.5). Thus, the reported results agree with modern work reporting relatively high CN for such biodiesel fuels. A later paper by another author reported the auto-ignition temperature of various alkyl esters of palm oil fatty acids [25]. In more recent times, the use of methyl esters of sunflower oil to reduce the viscosity of vegetable oil was reported at several technical conferences in 1980 and 1981 [26] and marks the beginning of the rediscovery and eventual commercialization of biodiesel. In a similar manner, palm oil was often considered as a source of diesel fuel in the historic studies [27]. Although, the diversity of oils and fats as sources of diesel fuel had been investigated and reported, however, an important aspect again today and striving for energy independence were reflected in other historic investigations [28]. Vegetable oils were also used as emergency fuels and for other purposes during World War II. [29]. For example, Brazil prohibited the export of cottonseed oil so that it could be substituted for imported diesel fuel. Reduced imports of liquid fuel were also reported in Argentina, necessitating the commercial exploitation of vegetable oils [28]. China produced diesel fuel, lubricating oils, gasoline and kerosene, the latter two by a cracking process and other vegetable oils. However, the exigencies of the war caused hasty installation of cracking plants based on fragmentary data. Researchers in India, prompted by the events of World War II, extended their investigations on 10 vegetable oils for development as domestic fuels. Work on vegetable oils as diesel fuel ceased in India when petroleum-based diesel fuel again became easily available at low cost [29]. The Japanese battleship *Yamato* reportedly used edible refined soya bean oil as bunker fuel [30]. Concerns about the rising use of petroleum fuels and the possibility of resultant fuel shortages in the United States in the years after World War II played a role in inspiring a "dual fuel" project at the Ohio State University, during which cottonseed oil, corn oil and blends with conventional diesel fuel were investigated [28]. In modern times, biodiesel is derived or has been reported to be producible from many different sources, including vegetable oils, animal fats, used frying oils and even soap stock [31]. Generally, factors such as geography, climate, and economics determine which vegetable oil is of greatest interest for potential use in biodiesel fuels [32]. Thus, in the United States, soybean oil is considered to be a prime feedstock; in Europe, it is rapeseed (canola) oil, and in tropical countries, it is palm oil. As noted above, different feed stocks were investigated in the historic times. These included palm oil, soybean oil, cottonseed oil, groudnut oil, castor oil and a few less common oils such as babassu and crude raisin seed oil; non-vegetable sources such as industrial tallow and even fish oils were also investigated [26]. Walton, (1938) summarized results on 20 vegetable oils (castor, grapeseed, maize, camelina, pumpkinseed, beechnut, rapeseed, lupin, pea, poppyseed, groudnut, hemp, linseed, chestnut, sunflower seed, palm, olive, soybean, cottonseed, and shea-butter). He also pointed out that "at the moment the source of supply of fuels is in a few hands, the operator has little or no control over prices or qualities and it seems unfortunate that at this date, as with the petrol engine, the engine has to be designed to suit the fuel whereas strictly speaking, the reverse should be the case, that is, obtaining the fuel and refining it to meet the design of an ideal engine" [26]. Although, environmental aspects played virtually no role in promoting the use of vegetable oils as fuel in historic times and no emissions studies were conducted, it is still worthwhile to note some allusions to this subject from that time; (i) "In case further development of vegetable oils as fuel proves practicable, it will simplify the fuel problems of many tropical localities remote from mineral fuel and where the use of wood entails much extra labor and other difficulties connected with the various heating capacities of the wood's use, to say nothing of the risk of indiscriminate deforestation" [33]. (ii) "It might be advisable to mention, at this juncture, that, owing to the altered combustion characteristics, the exhaust with all these oils is invariably quite clean and the characteristic diesel knock is virtually eliminated" [26]. (iii) Observations by other authors included: "invisible" or "slightly smoky" exhausts when running an engine on palm oil [34]; clearer exhaust gases; in the case of use of fish oils as diesel fuels, the exhaust was described as colorless and practically odorless [35]. These visual observations of yesterday have been confirmed in modern times for biodiesel fuel. Numerous recent studies showed that biodiesel fuel reduces most exhaust emissions [28, 31].

1.2. Production and processing of biodiesel

The production process of biodiesl involve combination of Alcohol, oil and catalyst in an agitated reactor, approx. at 60ºC during 1hr. Smaller plants often use batch mode reactors, but larger plants use continuous flows processes involving continuous stirred-tank reactors or plug flow reactors. Biodiesel as a renewable source of energy can be produced from different sources such as lipids, animal fats, vegetable oils and waste vegetable oils (Table 2). It has been reported that the production of biodiesel feedstock is sensitive [27, 36-37]. For instance, the cost of biodiesel can be estimated based on assumption regarding production volume, feedstock and chemical technology, hence feedstock cost comprises a substantial portion of overall biodiesel cost [38, 39]. Production of biodiesel from fats and oils is motivated by the fact that fat and oils contain a glycerol molecule bonded to three fatty acid chains, this structure is called a triglyceride and it is the major component of the oil. In general, biodiesel feedstock can be categorized into three groups: Pure vegetable oils (edible or non-edible), animal fats and waste cooking oils. Pure vegetable oil is also utilized for the production of biodiesel due to the fact that the oil composition from vegetable crops is pure [40]. Most liquid plant oils are composed predominantly 18-carbon fatty acids. Oils, which tend to be a liquid at room temperature, tend to have more double-bonds than fats. The major oils of this group are: rapeseed oil (canola), soybean oil, palm oil, sunflower oil, corn oil, groundnut (groudnut) oil, coconut oil, safflower oil, linseed oil, hemp, microalgae oil, Jatropha curcas and Pongamia pinnata [41]. Fats, which tend to be solid at room temperature, can also be used as a feedstock for the production of biodiesel. Fats have shorter carbon fatty acids and tend to have fewer double bonds than oils. These include: sheep tallow, yellow grease, beef, and poultry oil etc. Also used in the production of biodiesel is the waste cooking oils. These are also denominated in literatures as waste fried oils (WFO). It include: used rapeseed oil, used sunflower oil, used soybean oil, used cottonseed oil, used olive oil. The use of waste oil for the production of biodiesel has been identified as economical method, due to the availability of waste cooking oil, however processing and gathering of the waste cooking oil militate against its adoption. Commercial availability of biodiesel is therefore influence by the feedstock availability, which can be achieved in Nigeria through groundnut oil with production rate of 200 000 metric tonne per year.

S/N	Chemical Name	Fatty acid Methyl Ester
1	Chemical Formula Range	C_{14}-C_{24} methyl esters
2	Kinematic Viscosity Range	3,3- 5,2
3	Density Range	860-894
4	Boiling point Range (K) Flash Point Range (K)	>475 430-455
5	Distillation Range (K)	470-600
6	Vapor Pressure (mmHg at 295K)	<5
7	Solubility in water	Insoluble in water
8	Physical appearance	Light to dark yellow transparent liquid

Table 1. Physical Biodiesel Properties (Shay, 1993)

1.3. Availability and properties of groudnut oil

Groudnut oil (*arachis* oil) is an organic material oil derived from groudnuts and it has been noted to have the aroma and taste of its parent legume. It is often used in Chinese, South Asian and Southeast Asian cuisine as much as olive oil is used in the Mediterranean. Groudnut oil is appreciated for its high smoke point relative to many other cooking oils. Its major component fatty acids are oleic acid (46.8% as olein), linoleic acid (33.4% as linolein), and palmitic acid (10.0% as palmitin). The oil also contains some stearic acid, arachidic acid, arachidonic acid, behenic acid, lignoceric acid and other fatty acid [28, 29]. Vegetable oils are widely consumed domestically in Nigeria. It is used primarily as a cooking and salad oil. Studies have shown that groundnut oil contains much potassium than sodium and is a good source for calcium, phosphorus and magnesium. It also contains thiamin, vitamin E, selenium, zinc and arginine [42]. Findings have demonstrated that diets high in groudnut oil are as effective as olive oil in preventing heart disease and are heart healthy than very low fat diets [42]. Groudnut oil is of high quality and can withstand higher temperatures without burning or breaking down, has neutral flavour and odour, does not absorb odour from other foods and all these properties thus makes it the most preferred oil in Northern Nigeria [32]. The nutritional values of the groudnut oil are however affected by the method and period of storage, which consequently affect the acceptability of these oils. In Northern Nigeria, women usually do extraction of groudnut oil locally. The women extract the oil to generate substantial income to support their domestic needs with little or no consideration given to groudnut species or the physicochemical properties of the oil. Most women rely on availability rather than quality. Groudnut is mainly grown in the Northern part of Nigeria and the oil is readily available in all parts of Nigeria in large quantities. This is because it has been reported that Nigeria possesses land area of 923,768 km^2 arable land constituting about 56% and vegetation ranging from the Sahel savanna in the extreme North to swamp forest in the south [32]. Therefore most parts of Nigeria are suitable for biofuel crop cultivation. Although an edible oil, its use as a potential feedstock for biodiesel production may not likely compete with other crops grown for food and commercial cooking oil products. Nigerian is ranked 3rd in the world for the production of groundnut oil as shown in Table 3 (USDA, 2009)[43]. The annual production of groundnut in Nigeria is shown in Figure 1. The reduction in the production of groundnut oil between the year 1976 and 1994 can be attributed to the diversion of Nigerian economy from agricultural oil exploitation economy due to discovery of oil in Nigeria. The recent interest (from 1999 till date) in the production of groundnut can be attributed to the Nigerian government commitment to the production of alternative renewable energy sources. Biodiesel has been identified as the alternative energy in the country with groundnut oil as the perfect feedstock for its production without affecting the oil supply for consumption purpose.

S/No	Vegetable oil	Production (billion/year)	Animal Fats	Production (pounds/year)
1.	Soybean	18.340	Edible tallow	1.625
2.	Groudnut	0.220	Inedible tallow	3.859
3.	Cotton seed	1.000	Lard and Grease	1.306
4.	Sunflower	1.010	Yellow Grease	2.633
5.	Corn	2.420	Poultry Fat	2.215
6.	Others	0.669		
	TOTAL	**23.659**	**TOTAL**	**1.638**

Table 2. Annual Oil Productions of Some Biodiesel Sources [44]

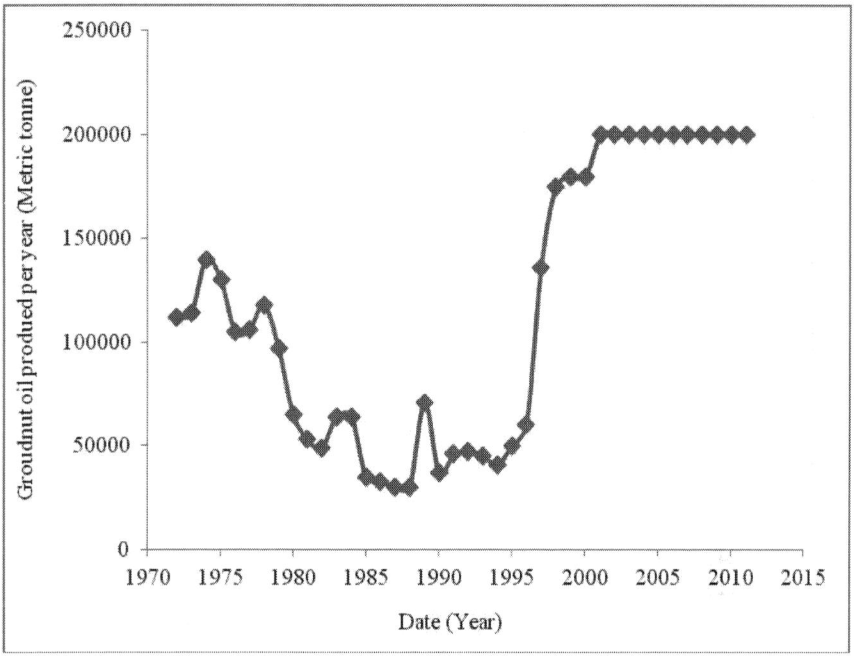

Figure 1. Nigeria's Groundnut (Peanut) Oil Production per Year (United States Department of Agriculture, 2011)

S/No	Country	Production Million metric tons
1.	People's Republic of China	14.30
2.	India	6.25
3.	United States	2.34
4.	Nigeria	1.55
5.	Indonesia	1.25
6.	Myanmar	1.00
7.	Sudan	0.85
8.	Senegal	0.71
9.	Argentina	0.58
10.	Vietnam	0.50
11.	TOTAL	34.43

Table 3. World's Top Ten Producers of Groundnut (Groudnut) [43].

It has been estimated that in the year 2000, approximately 34 million Mt of groundnuts were produced world wide of which 15 million Mt were produced in China, 6 million Mt in India, 2 million Mt in Nigeria, 1.5 million Mt in United States of America and the rest mostly in other countries [45]. Groundnuts are important component of Nigerian diet and about 5 percent of the estimated 58.9 g of crude protein available per head per day is contributed by groundnut [45]. In most of the developing countries it provides high-quality cooking oil and is an important source of protein for both human and animal diet and also provides much needed foreign exchange by exporting the kernels and cake. Hence, the utilization of groundnut oil for biodiesel production without affecting the supply of the vegetable oil for consumption.

2. Material and methodology

The feedstock used in this work for the production of biodiesel is refined groundnut oil produced by Grand Cereals Mill Zawan, Plateau state, Nigeria. This product was purchased at a local super market in Minna Niger State, Nigeria. The entire chemicals used in this study are of analytical grade (98-99.5%). They include Carbon tetra chloride (Analar, BDH), Wij's solution (Mixture of glacial acetate, iodine trichloride and carbon tetralchloride) (Hopkins and Williams, London), potasium hydroxide solution (Analar, BDH), potassium hydroxide pellet (Burgoyne & co, India), petroleum ether (Analar, BDH), potassium iodide solution (M&B, England), sodium thisosulphate (M&B, England), hydrochloric acid (Analar, BDH) and potassium iodide pellet (M&B, England). The equipments used are pH meter, distillation apparatus, viscometer, thermostatic hot plate, sulphur in oil analyzer, magnetic stirrer, digital weighing balance, thermometer, petri dish, pippete, separating funnel, burete, oil test centrifuge, pycometer bottles, Abbe refractometer, flash point tester, oven and aneline point teseter.

Production of biodiesel from groundnut oil was carried out in accordance to the experimental procedure reported by Hill (2002) [45]. Methoxide was first prepared in a suitable container by dissolving sodium hydroxide in methanol. This typically consists of 50% methanol by volume of oil and 0.25 grams sodium hydroxide for lower limit and 0.50grams sodium hydroxide for upper limit mixed together well so that the sodium hydroxide is completely dissolved in the methanol in the ratio of 1:4 for lower limit and 1:6 for upper limit respectively. Transesterification reactions were carried out in a reaction vessel (250ml flask). The reactor was filled with 50g refined groundnut oil sample and preheated to the selected reaction temperature (40^{0}C and 60^{0}C); the methoxide solution was then added on top of the oil. The mixture was agitated vigorously for about 30 seconds and then placed on a magnetic stirrer for stirring to continue for the reaction time (45 and 90 minutes respectively). The transesterification was carried out on the hot plate to keep the reaction temperature constant, after reaction time was reached, the reaction mixture was then poured into a separating funnel and allowed to settle gravitationally for about 8-20hrs. After settling, a lighter colored biodiesel on top of a layer of darker glycerin was observed. The bottom layer (glycerol) was drained off and the top layer (methyl esters) was collected in a clean beaker.

Characterization of the feedstock (refined groundnut oil) is very important for proper understanding of its chemical composition, as well as physico-chemical properties to ease comparison with standard and possibly establish the level of purity of the material relative to using it for biodiesel production. Properties of the feedstock tested for are the specific gravity (or density), flash point, kinematic viscosity, acid value, iodine value, sulphur content, moisture content, free fatty acid and acid value. Determination of these propertics was carried out using the experimental description reported by Onwuka (2005) [46]. Next to the prelimenary production of biodiesel from groudnut oil was the process optimization by investigating the effects of various parameters on the yield of biodiesel from groudnut oil.

Effects of operating parameters such as temperature, time of reaction, and catalyst concentration as well as oil-methanol mass (or mole) ratio were investigated. A 2^4 factorial experimental design was used to carry out the study. A 2^4 factorial experimental design was used to determine the optimum conditions for biodiesel production from groundnut oil, four variables were studied at both high and low levels. The expected response is biodiesel (methyl ester) yield. The low level of methanol: oil mass ratio was 6:1 and the high level was 4:1. The low level of catalyst concentration chosen was 0.5% and the high level was 1.0% NaOH catalyst by weight of groundnut oil. The low level of temperature was chosen as 40 °C and the high level was chosen at 60 °C, which was determined by the boiling point of methanol. The reaction time chosen for the lower level was 45 minutes and 90 minutes for the higher level. Table 4 shows the experimental matrix for the 2^4factominrial designs (4factors, each run at two levels). The biodiesel produced was then analyzed to determine the specific gravity, flash point, cloud point, kinematic viscosity, pour point, centane number, acid value, bottom water and sediment, sulphur content, ash content and distillation characteristics. All experimental analysis were conducted in triplicate and the results reported are the average values with average deviation of ±0.0015

Run	Temperature (°C)	Methanol : oil ratio (wt/wt)	Catalyst Conc. (wt %)	Reaction time (min)
1	40	1:4	0.5	45
2	60	1:4	0.5	45
3	40	1:6	0.5	45
4	60	1:6	0.5	45
5	40	1:4	1.0	45
6	60	1:4	1.0	45
7	40	1:6	1.0	45
8	60	1:6	1.0	45
9	40	1:4	0.5	90
10	60	1:4	0.5	90
11	40	1:6	0.5	90
12	60	1:6	0.5	90
13	40	1:4	1.0	90
14	60	1:4	1.0	90
15	40	1:6	1.0	90
16	60	1:6	1.0	90

Table 4. 2^4 Factorial design

3. Results and discussion of results

Refined groundnut oil that was used as the feedstock for the production of biodiesel was characterized to determine the necessary properties that are very essential to the quality and quantity of the final product. Table 5 summarized the results obtained on the qualities of the refined groundnut. Presented in Table 6 is the results obtained on the influence of the operating parameters on the yield of biodiesel, while Table 7 represent the properties of the biodiesel produced from refined groundnut oil. Also presented is the statistical analysis of the results obtained to illustrate the effect of each of the operating parameters on the yield of the biodiesel from the refined groundnut oil.

Table 5 summarized the results of characterization carried out on the sample of refined groundnut oil used and the results were compared with standard and available literature. The main properties determined include saponification value, iodine value, acid value, free fatty acid, peroxide value, refractive index, kinematic viscosity, pH, unsaponifiable matter and the moisture content, results obtained are hereby presented.

The saponification value of as reported by Al-Zahrani (2005) [48] is a measure of the tendency of oil to form soap during transesterification reaction (i.e. define as the number in milligram (mg) of KOH required to saponified 1g of the sample fat).The saponification value obtained for sample of refined groundnut oil was 186.53, which is lower compared to 188-195 recommended by AOCS standard(1998) [46], but higher than 148.67 reported for the variety used by Ibeto *et al.* (2011) [47].The high saponification values indicate the presence of high percentage of fatty acids which might lead to soap formation and hence low biodiesel yield [48]. Unsaponifiable matter are those substances frequently found dissolved in fatty acids and drying oils which cannot be saponified by caustic treatment, but which are soluble in normal fat solvents. The unsaponifiable value for the oil sample used oil was 0.29% which is very reasonable compared to 1% maximum recommended by AOCS standards as presented in Table 5. Iodine Value (IV) is a value of the amount of iodine, measured in grams, absorbed by 100 grams of given oil. The iodine value is expressed in grams of iodine for the amount of halogens linked with 100g test sample, and is used as degree of unsaturated bond of fats and oils. Iodine values are used to classify oils as either drying oil (>130), semi-drying oil (115-130) and non-drying oil (<115) (Othmer, 2011) [49]. The oil used in the present study has iodine value of 94.16gI_2/100g, although higher than 89.46gI_2/100greported for the groundnut oil sample used by Ibeto *et al.* (2011) [47] but falls within the 84-100gI_2/100g range recommended by AOCS standard of 1998. Based on this value, the present oil in use can be classified as non-drying oil. Also measured is the acid value (AV) of the oil which is an important indicator of vegetable oil quality. Vegetable oils with high acid value are classified as inedible while those with low acid value are classified as edible oils [48]. The obtained acid value for the present oil sample was 2.174mgKOH/g, a value that falls within 0.72-3.0mgKOH/g reported in the AOCS standard (1998) [46] but lower when compare to 2.61mgKOH/g obtained by Ibeto *et al* (2011) [47]. However, both reported values of the present study and that of Ibeto *et al.* (2011) [47] indicated the oil samples are edible type. Results as presented in Table 5 also indicate that the Free Fatty Acid (FFA) value of the refined groundnut oil sample was 0.05%, and was found to fall within the AOCS standard having its FFA value as <1% (AOCS, 1998) . Literature also revealed that for groundnut oil to be fit for biodiesel production its free fatty acid content should be <2%, this is to enable efficient conversion of the oil into biodiesel (Anggraini and Wiederwertung, 1999) [50]. Ibeto *et al* (2011) obtained a free fatty acid value of 1.31mgKOH/g, translating to 0.065%which was a little bit higher than 0.05% reported in this study.

S/No	Properties	Unit	Experimental values	AOCS Standard value (1998)	Ibeto *et al.*, (2011)
1.	Specific gravity (S.G)	-	0.91	0.910-0.915	0.93
2.	Kinematic Viscosity at 25°C	$mm^2\ sec^{-1}$	38.00	-	32.66
3.	Saponification value (S.V)	mgKOH/g	186.53	188-195	148.67
4.	Moisture content	%	0.05	-	0.09
5.	Iodine value (I.V)	$gI_2/100g$	94.16	84-100	89.46
6.	Unsaponifiable matter	%	0.29	1% max	-
7.	pH		5.56	-	-
8.	Refractive index (R.I) at 25°C	-	1.467	1.467-1.470	1.463
9.	Acid value (A.V)	mgKOH/g	2.174	0.72-3.0	2.61
10.	Free fatty acid (FFA)	%	0.05	<1%	1.31
11.	Peroxide value	mEq/kg	0.059	-	22.25

Table 5. Measured Physiochemical Properties of Refined groundnut oil.

Kinematic viscosity refers to the thickness of the oil and its value is meaningless unless accompanied by the temperature at which it is determined. The purpose of transesterified vegetable oils and animal fats into alkyl esters (biodiesel) is to reduce the kinematic viscosity because high viscosity tends to form larger droplets on injection which can cause poor combustion, increased exhaust smoke and emissions. Hence, kinematic viscosity is key factor in checking suitability of a feedstock for the production of biodiesel. Result obtained as presented in Table 5 shown that the kinematic viscosity of the refined groundnut utilized in this study for the production of biodiesel is $38mm^2/sec$ at 25°C, which is much higher compared to 32.66 mm^2/sec obtained by Ibeto *et al* (2011). While the specific gravity of the refined groundnut oil was found to be 0.91, which falls within the range of AOCS specification (0.910-0.915) and lower than 0.93 reported by Ibeto *et al.* (2011). The peroxide value obtained for the refined groundnut oil was 0.059mEq/kg which was very low compared to 22.25 mEq/kg obtained by Ibeto *et al* (2011). The value obtained is considered reasonable, because literature suggest that the peroxide values of fresh oils should be less than 10 millequivalents /kg, peroxide values between 30 and 40 millequivalents/kg will results into rancidity of the oil. The moisture content in the oil sample as shown in Table 5 was found to be 0.05% and much lower compared to 0.09%reported for the oil sample used by Ibeto *et al* (2011).The 0.05% value obtained agreed with the value recommended by ASTM (D 6571) and falls within the <3% moisture content suggested for all raw materials to be used in the production of biodiesel (Freedman *et al.*, 1984). It has been reported that moisture content greater than <3% will decrease the efficiency of the transesterification reaction due to possibility deactivation of catalyst active sites and soap formation (Freedman *et al.*, 1984). The result obtained on the refractive index of the oil sample at 25^{0}C was 1.467 which was within the recommended AOCS standard index of 1.467-1.470 and also compared well with 1.463 reported by Ibeto *et al.* (2011). It can be inferred from the various analysis conducted on the refined groundnut oil sample that the oil sample compare favorably with the standard and literature value, hence the oil sample utilized in this study should produced biodiesel with high yield and quality.

The effect of reaction parameters on the yield of biodiesel produced from refined groundnut oil was investigated using a 2^4 factorial design. Reaction parameters considered include mole ratio of methanol to refined groundnut oil, reaction temperature, catalyst concentration and reaction time. Each of these parameters was considered at two specified levels and the summary of results obtained was presented in Table 6. As shown in Table 6, the best biodiesel yield of 99% was obtained under the optimal conditions of 1:6 (oil to methanol) molar ratio, 0.5% catalyst concentration at 40°C and for a reaction time of 90 minutes. This value was better than 88% and 79% reported by Galadima *et al.*(2008) and Ibeto *et al* .(2011) respectively. The better ester production yield obtained could be attributed to different reaction parameters used in conducting the experiment, different alcohol and alkali catalyst and possible variation in the specie of groundnut oil used which is attested to the values obtained for the characterization of the oil (Table 5).

One of the most important parameters that affects biodiesel yield is the oil: alcohol ratio (in this case refined groundnut oil to methanol) mole ratio [51]. In order to evaluate its effect on biodiesel yield, transesterification was conducted at two different molar ratios of 1:4 and 1:6. Though, the stoichiometric ratio for transesterification requires one (1) mole of triglyceride and three (3) moles of alcohol to yield three moles of fatty acid alkyl esters and one mole of glycerol. However, transesterification is an equilibrium reaction in which a large excess of alcohol is required to drive the reaction to the product side. In this work, two levels of Groundnut Oil to MeOH molar ratio were used, 1:4 for the low level and 1:6 for the high level. The yields ranging from 94-99% were obtained at a molar ratio of 1:6. It can be seen from Figure 2 that the biodiesel yield increased for both 0.5 wt% and 1.0 wt% catalyst concentration as the molar ratio was increased from 1:4 to 1:6 i.e. from 96 to 99% for 0.5wt% and from 80 to 92% for 1.0wt% catalyst weight respectively. While temperature and time were constant and this achieved experimental result is in agreement with Freedman *et al* (1984) [52]. Since transesterification is an equilibrium process, lower oil to methanol molar ratio may result in an incomplete transesterification and thus increasing the molar ratio which will shift the reaction to the ester formation direction. However when the molar ratio

Run	Temperature	Oil: Methanol mole ratio	Catalyst weight (wt %)	Reaction time	Production yield (wt %)	Methyl ester concentration (wt %)	Methyl ester yield (wt %)
1	40	1:4	0.50	45	96	97.92	94
2	60	1:4	0.50	45	90	80.00	72
3	40	1:6	0.50	45	96	97.92	94
4	60	1:6	0.50	45	98	97.96	96
5	40	1:4	1.00	45	94	97.87	92
6	60	1:4	1.00	45	76	92.11	70
7	40	1:6	1.00	45	96	93.75	90
8	60	1:6	1.00	45	96	95.83	92
9	40	1:4	0.50	90	100	96.00	96
10	60	1:4	0.50	90	98	97.96	96
11	40	1:6	0.50	90	101	98.02	99
12	60	1:6	0.50	90	98	97.96	96
13	40	1:4	1.00	90	82	97.56	80
14	60	1:4	1.00	90	80	95.00	76
15	40	1:6	1.00	90	94	97.87	92
16	60	1:6	1.00	90	92	97.83	90

Table 6. Biodiesel Yield at Different Conditions

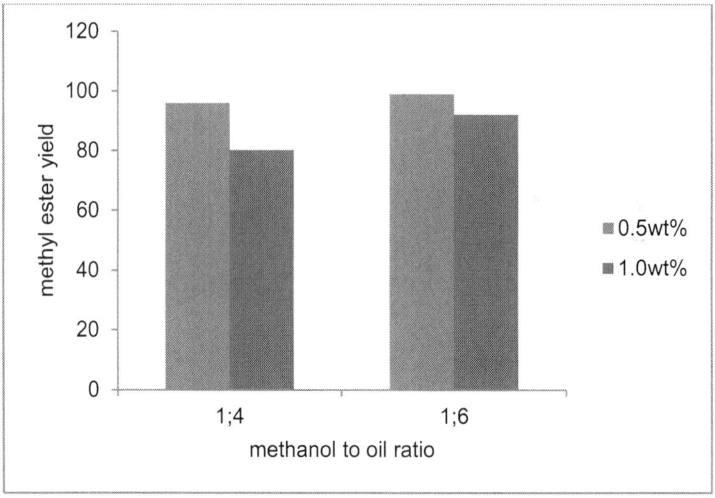

Figure 2. Effect of methanol to oil ratio on methyl ester yield at 40°C and 90 minutes reaction time.

is set too high, the excessive alcohol may favor conversion of triglycerides to diglycerides and then monoglycerides and a slight recombination of esters and glycerol to monoglycerides because their concentrations keep increasing during the course of the reaction. The detailed analysis of the results obtained on the effects of various parameters on the yield of biodiesel from groudnut oil are presented below.

The rate of reaction and biodiesel yield can be affected by reaction temperature, Anitha and Dawn, (2010) in their work varied the temperature between 40-65^{0}C and concluded that the primary advantage of a high temperature is shorter reaction time. Transesterification can occur at different temperatures depending on the oil used. Several researchers found that the temperature increase influences the reaction in a positive manner (Ma and Hanna, 1990; Freedman *et al.*, 1984; Canakci and Van Gerpen, 1999). Though, it has been reported that the best temperature for the reaction is 60°C but depending on the type of catalyst, different temperatures will give different degrees of conversion. In this work, the two levels of temperature used were 40°C for the low level and 60°C for the high level. The best yields were obtained at the low temperature of 40°C as presented in Figure 3. This pattern of results can be attribute to the fact at a high temperature of 60°C most of the alcohol in the biodiesel would evaporate during transesterification reaction leading to a lower yield Anitha and Dawn (2010), also confirmed this. The figure also confirmed that as the temperature increase from 40^{0}C to 60^{0}C the methyl ester yield decreased from 94 to 92% for 0.5wt % catalyst weight and from 92 to 70%for 1.0wt% catalyst weight respectively.

Analysis of result obtained as presented in Table 5 shows that increase in time does not implies increase in yield as suggested by Yusuf and Sirajo (2009). For instance, at 40°C using 1:4 mole ratios with 1wt% catalyst for 45min gave a yield of 92% while the yield decreased to 80% at the same condition but with 90min reaction time. In this study, two levels of reaction time were used i.e. 45 minutes for the low level and 90 minutes for the high level. The optimum yield (99%) was achieved at 90minutes and this observation is in agreement with Freedman's work and Ahmad *et al.*, (2009). Figure 4 which also depict the effect of time on the methyl ester yield indicate that the maximum methyl ester yield at 45minutes was

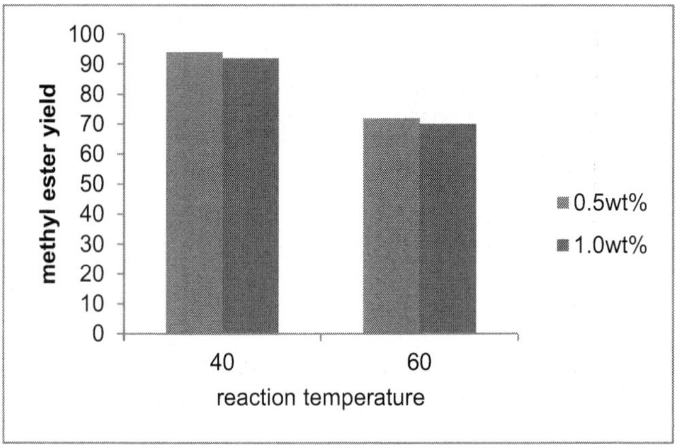

Figure 3. Effect of reaction temperature on methyl ester yield at 40^{0}C at 45 minutes reaction time.

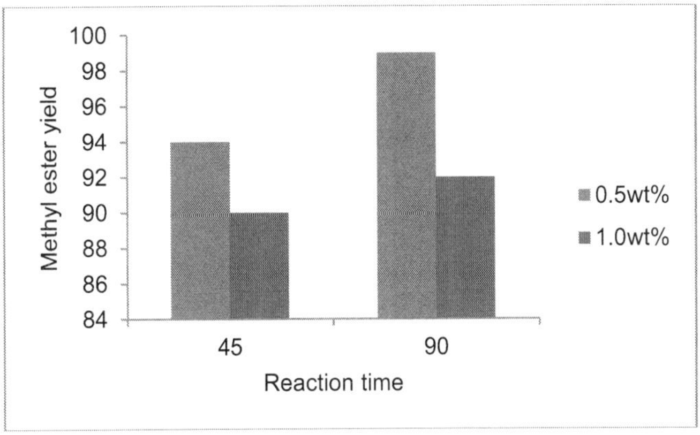

Figure 4. Effect of reaction time on methyl Ester Yield at 40^{0}C using 1:6 methanols to oil ratio.

96% which is less than the value obtained at 90 minutes (99%) at same reaction conditions. This indicates that there is increase in biodiesel yield as reaction time increases although this could be as a result of varying other variables too.

Also investigated is the effect of catalyst on the yield of the biodiesel from grudnut oil. It has been reported that as the catalyst concentration increases, the conversion of triglyceride and the yield of biodiesel also increase. This phenomenon was attributed to an increase in the availability and number of catalytically active sites (Anitha and Dawn, 2010). This is because an insufficient amount of catalysts will result in an incomplete conversion of the triglycerides into the fatty acid esters (Anitha and Dawn, 2010). Two levels of catalyst concentration (NaOH) were used in this work, 0.5wt% for the lower and 1.0wt% for the higher level. From the obtained experimental result shown in Table 6 and Figure 5, it was observed that the best yield which was 99% was obtained at 0.5wt% NaOH, but at 40^{0}C, 1:6 and a reaction time of 90minutes. Although the smaller the catalyst concentration, the lower the yield due to insufficient amount of catalyst to catalyze the reaction to a completion stage but this result was achieved due to the excess of alcohol used in this condition. It can be seen from Figure 5 also that the methyl ester yield is higher at catalyst concentration of 0.5 wt% than at 1.0wt%. Also, it was observed that at conditions other than the optimal conditions of 1:6 molar ratio, 60°C and 90 minutes reaction time, 1.0 wt% catalyst weight gave a lower yield in most run than the 0.5 wt% for same conditions. Most researchers said that an increase in catalyst concentration increases the methyl ester yield (Anitha and Dawn, 2010; Yusuf and Sirajo, 2009) but from this research, higher catalyst concentration gave lower yield, as shown in Figure 5.

Biodiesel produced from refined groundnut oil using alkali catalyzed transesterification process with sodium hydroxide (NaOH) as catalyst and methanol as alcohol was characterized to determine its suitability as alternative to petrol diesel. Results obtained on the various analysis conducted are presented in Table 7.

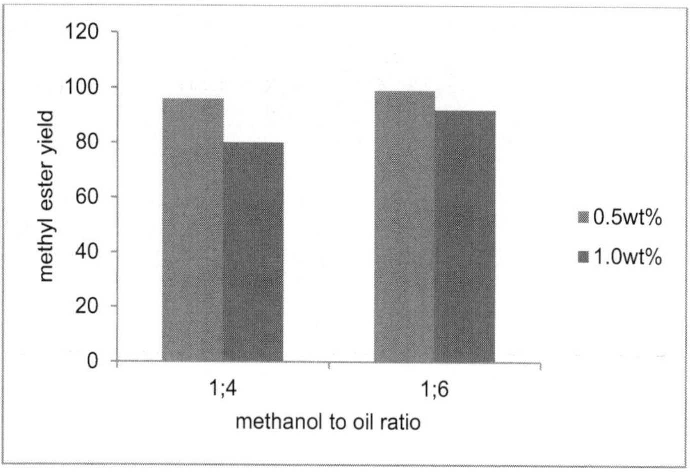

Figure 5. Effect of Catalyst Concentration on Ethyl Ester Yield.

Results as presented in Table 7 indicate that the kinematic viscosity of the biodiesel produced was determined as 5.86 mm^2/ s, a value that falls within the range of 1.9-6.0mm^2/s recommended by ASTM (DS 6751) and a little higher than 5.16mm^2/s reported by Ibeto *et al.* (2011). However, the value obtained shows that the transesterification reaction has successfully reduced the viscosity of oil from38mm^2/s to 5.86 mm^2/s and as such increases its tendency to flow. Since kinematic viscosity described the resistance of a fluid to flow under gravity and it is important in determining optimum handling, storage, and operational conditions. High viscosity can cause fuel flow problems and lead to stall out or fuel pump failure [54]. Also presented in Table 7 is the results obtained for the bottom sediment and water content of the biofuel produced. Presence of water and sediment in biodiesel beyond standard limit will results in poor ignition, filter clogging, and fuel pump problems, sediment in storage tanks can obstruct the flow of fuel from the tank to a combustor and presence of water in middle distillate fuels can cause corrosion and growth of microorganisms. According to ASTM standard, a maximum water content of 0.05%vol fuels is acceptable, while the biodiesel produced contain trace quantity of water and sediment. This is similar to the results reported by Ibeto *et al.* (2011) and better than 0.02% water and sediment reported by Galadima *et al.* (2008) [53]. This result gave an indication that the biodiesel is clean and properly dried. Sulphur content of the biodiesel produced was measured and the result obtained is as shown in Table 7. Pure biodiesel (B100) is expected to contain very low quantities of sulfur in order to retain its traditional name of sulphur-free fuel. The sulphur content of the biodiesel produced was found to be 0.025 wt% (Table 7) and this is low when compared to ASTM maximum limit of 0.050wt%. It can be inferred that the fuel produced is "Sulphur free". Also presented in Table 7 is the distillation characteristic of the produced biodiesel, though for a pure substance the boiling point is a single temperature value. However, when a mixture of hydrocarbons exists in diesel fuel there is a range of boiling points for the different constituent chemical specie, the boiling point of the biodiesel

Properties	Units	Experimental values	ASTM standard biodiesel (ASTM D 6751)	ASTM standard Petrol-diesel (ASTM D 975)	Ibeto *et al.*, (2011)	Galadima *et al.*, (2008)
Specific gravity	_	0.88	-	0.95max	0.88	0.84
Kinematic Viscosity at 40^{0}C	mm^2 / s	5.86	1.9-6.0	1.9-4.1	5.16	-
Flash point	^{0}C	170	130min	60-80	202°C(395.6^{o}F)	-
Cloud point	^{0}F	57	-	-15 to 5	-	-
Sulphur content (by X-Ray)	% mass	0.025	0.05max	0.05max	-	-
Distillation I.B.P	^{0}C	350	-	-	-	-
10%	^{0}C	355	-	70max	-	-
50%	^{0}C	365	-	125max	-	-
80%	^{0}C	>400	-	180max	-	-
E.B.P	^{0}C	_	360max			
Total recovery	-	97%	-	-	-	-
Cetane index	_	49.56	47min	40-55	-	-
Ash content	% mass	0.016	0.02	-	-	
Free glycerine	% mass	0.018	0.02max	0.02	-	-
Total Glycerine	% mass	0.240	0.24max	0.24	-	-
Acid value	mg KOH/g	0.210	0.5max	-	4.96	0.45
Water by distillation	% vol	Trace	0.05max	0.5max	Trace	-
Basic water and Sediment	% vol	Trace	0.05max	0.5max	Trace	0.02

Table 7. Measured Properties of biodiesel produced

produced was taken at its initial boiling point (IBP), at 10%, 50%, and 80% and at its end boiling point, the values obtained for each of them were 350^{0}C, 355^{0}C, 365^{0}C and >400^{0}C respectively these values were found to be much higher than the ASTM D 6751 due to the fact that a vacuum distillation apparatus was used. An acid value of the produced biodiesel was also measured and the result is presented in Table 7. The acid number correlates to the fuel's long-term stability and corrosiveness, the smaller the acid value, the higher the quality of the fuel. The acid value of the biodiesel produced was found to be 0.210mgKOH/g, a value lower compared to the maximum recommended standard value of 0.5mgKOH/g (ASTM D 6751) and better than 4.96 and 0.45mgKOH/g reported by Ibeto *et al.* (2011) and Galadima *et al.* (2008) respectively. The Cetane number relates to the readiness of the fuel to self-ignite when exposed to the high temperatures and pressure in the diesel engine combustion chamber. The cetane number affects a number of engine performance parameters such as combustion, stability, drivability, white smoke, noise and emission of CO. The number is also indicative of the relative fuel stability [55]. The cetane number of the biodiesel produced was calculated as 49.5, a value above the ASTM acceptable minimum of 47(as expected) which indicates ignition delay and better ignition properties [55].

Also measure and presented in the table of result is the flash point of the produced biodiesel, which is described as a tendency of a sample of fuel to form a flammable mixture with air. The flash point of the biodiesel produced was found to be 170°C which is greater than the minimum acceptable value of 130°C specified by ASTM D 6751 (Table 7). Cloud point is a fuel property that is particularly important for the low temperature operability of diesel fuel. Initially, cooling temperatures cause the formation of solid wax crystal nuclei that are submicron in scale and invisible to the human eye. The temperature at which crystal agglomeration is extensive enough to prevent free pouring of fluid is called its pour point [56]. In this work, the cloud point of the biodiesel was determined as 20⁰F and is in accordance with the work of Dunn *et al.,* (1996) [57]. These features have implications on the use of biodiesel in cold weather applications. Free and total glycerin values are indicators of incomplete esterification reactions and predictors of excessive carbon deposits in the engine it shows the mono, di and triglycerides molecule in the biodiesel. The total and free glycerin determined for the biodiesel produced was 0.018wt% and 0.24wt% respectively, both values were seen to be within the acceptable standard of ASTM D 6751 (Table 7) and also agreed with finding of Ibeto *et al.*(2011) except for slight difference in free glycerine which they obtained 0.02wt%.

The regression model of experimental data with respect to methyl ester conversion can be determined by least squares method. A statistical analysis was carried out on the experimental results obtained using analysis of variance (ANOVA). Effects of the four (4) factors and their interaction effect were estimated. The results of test of statistical significance in Table 8 shows that the four reaction parameters: molar ratio, temperature, time and catalyst concentration had different degrees of effects on the ethyl ester yield percentage with contributions of 11.5022%, 23.2275%, 14.5916% and 2.81534% respectively .

Table 8 shows that temperature had the highest effect of 8.4375 and the highest percentage contribution on methyl ester yield than other parameters having 23.23% contribution. Molar ratio showed the least effect of -5.9375 and a %contribution of 14.5916 on methyl ester yield. Catalyst concentration had an effect of -6.6875 with a percentage contribution of 14.5916%. The reaction time had an effect of 2.9375 but the least contribution on methyl ester yield with 2.81534%. It was also observed that the interaction effect between the factors were very significant. For instance, interactions between two factors such as temperature – molar ratio (A*B) had the highest effect of 6.0625 with the highest percentage contribution (11.9916%)

Term	Effect	SumSqr	% Contribtion
A-Molar Ratio	-5.9375	282.031	11.5022
B-Temperature	8.4375	569.531	23.2275
C-Catalyst Weight	-6.6875	357.781	14.5916
D-Reaction Time	2.9375	69.0313	2.81534
AB	6.0625	294.031	11.9916
AC	0.1875	0.28125	0.0114704
AD	4.5625	166.531	6.79174
BC	2.3125	42.7813	1.74477
BD	-2.0625	34.0313	1.38792
CD	-5.1875	215.281	8.77993
ABC	0.6875	3.78125	0.154213
ABD	-5.9375	282.031	11.5022
ACD	-0.8125	5.28125	0.215388
BCD	3.3125	87.7813	3.58003
ABCD	1.1875	11.2813	0.460089

Table 8. Effects of reaction parameters on methyl ester yield.

while temperature- reaction time (B*D) had the least effect of -2.0625 with the lowest percentage contribution of 1.388%. For interactions between three factors for instance, temperature-catalyst concentration-reaction time (B*C*D) had the highest effect of 3.3125 with a percentage contribution of 3.58003%, while molar ratio-catalyst concentration-reaction time (A*C*D) had the lowest effect of -0.8125 with a percentage contribution of 0.2154%. In concise, the molar ratio and catalyst weight were seen to have negative effect but temperature and reaction time were seen to have positive effects, while all interactions also had positive effects with the exception of (B*D), (C*D), (A*B*D) and (A*C*D) which had negative effects. However, the negativity of interactions is probably due to side reactions such as soaps formation (Vicente *et al.*, 1998). Table 10 present the variance analysis of the results.

The variance analysis of the results indicates that the Model F-value is 84.69 which implies that the model is significant as seen on Table6. this table shows the analysis of variance in which both the main effects(temperature, molar ratio of methanol-to-oil, rection time and catalyst weight), second order effect (AB, AD, BC, BD, CD), third order effect (ABD, BCD, ABC, ACD) and the fourth order effect(ABCD) are significant in the regression model but have no interaction effect. This implies that the values of "Prob > F" less than 0.0500 indicate model terms are significant. In this case A, B, C, D, AB, AD, BC, BD, CD, ABD, BCD, ABCD are significant model terms. Values greater than 0.1000 indicate the model terms are not significant. Stastitical analysis of the results also indicates that S.D. = 1.38; Mean = 89.47; C.V.% = 1.54; R^2 = 0.9876; R^2_{ADJ} = 0.9759; R^2_{PRED} = 0.9502; PRESS = 122.00; Adequate Precision = 28.168.

The "Predicted R-Squared" of 0.9502 is in reasonable agreement with the "Adjusted R-Squared" of 0.9759. "Adequate Precision" measures the signal to noise ratio. A ratio greater than 4 is desirable which shows that the ratio obtained 28.168 is adequate.

The analysis of variance showed that A,B,C,Dwere significant factors. The mathematical equation models for predicting average methyl ester yield are the final equation in terms of coded factors (Equation 1) and final equation in terms of actual factors (Equation 2)

$$\begin{aligned} Y = 89.47 - 2.97A + 4.22B - 3.34C + 1.47D + 3.03A*B + 0.094A*C + 2.28A*D + 1.16B*C \\ -1.03B*D - 2.59C*D + 0.34A*B*C - 2.97A*B*D \\ -0.41A*C + *D\ 1.66B*C*D + 0.59A*B*C*D \end{aligned} \quad (1)$$

$$\begin{aligned} Y = 453 - 95.00000A - 6.25000B - 76.00000C - 4.31111D + 1.62500AB + 34.00000AC \\ + 1.21111AD + 1.35000BC + 0.078889BD + 1.06667CD - 0.57500ABC - 0.021111ABD \\ - 0.60000ACD - 0.023333BCD + 0.010556ABCD. \end{aligned} \quad (2)$$

From Equations 1 and 2, all coefficients in the first and third order term have a minus sign which indicates that the %methyl ester conversion increases with these factors not conforming to the experimental result, while coefficients in the second and fourth order term have plus signs which indicates that the %methyl ester conversion increases with these factors conforming to the experimental result and also indiccates that there are also optimums within the experimental condition range.

4. Conclusion

Biodiesel was successfully produced from refined groundnut oil via alkali catalyzed transesterification using methanol, in the presence of NaOH. The produced biodiesel was characterized and from the results obtained, the methyl ester produced can be effectively used in a diesel engine since it meets the requirements of ASTM D 6751. Refined groundnut oil has also been proved to be a good feedstock for biodiesel as an optimum yield of 99% was obtained, and the obtained fuel properties such as kinematic viscosity (5.86 mm^2/s), flash point (170°F), cetane number (49.56), total sulphur content (0.025 wt%), e.t.c. after detailed characterization of the produced methyl esters of the oil which were within the standard ascribed by ASTM. Reaction temperature, reaction time, catalyst concentration and molar ratio of methanol to groundnut oil are the main variables affecting the methyl ester yield. These variables were optimized using a 2^4 factorial design while the agitation speed was kept constant at 200rpm and the optimum methyl ester yield (99%) was obtained at optimum reaction parameters of 40°C reaction temperature, 90 minutes reaction time, 0.5wt% catalyst concentration and methanol to oil molar ratio of 6:1.The regression models by least square method used to predict % methyl ester conversion from groundnut oil shows that the main effect, first order effect, second order effect, third order and fourth order effect are significant with no interaction effect, the coefficient of determination R^2 is 0.99. From the analysis of variance, it can be concluded that temperature had the highest effect and percentage contribution of 8.4375 and 23.2275% respectively to methyl ester yield from groundnut oil while molar ratio had the lowest effect (-5.9375) and a percentage contribution of 11.50%. It can also be concluded that from all the interaction effect of the four (4) reaction variables, A, B, C, D, AB, AD, BC, BD, CD, ABD, BCD, ABCD were significant model terms. A simple first degree polynomial regression model that can predict the methyl ester yield from groundnut oil has been developed and represented as:

$$Y = = 453 - 95.00000A - 6.25000B - 76.00000C - 4.31111D + 1.62500AB + 34.00000AC + 1.21111AD + 1.35000BC + 0.078889BD + 1.06667CD - 0.57500ABC - 0.021111ABD - 0.60000ACD - 0.023333BCD + 0.010556ABCD.$$

Based on the results of the various analysis conducted, it can be infered that, groundnut oil is a good feedstock for methyl ester production especially in Nigeria where it is cheap and readily available.

Author details

Jimoh A., Odigure J.O. and Odili U.C.
Department of Chemical Engineering, School of Engineering and Engineering Technology, Federal University of Technology, PMB 65 Minna, Niger State, Nigeria

Abdulkareem A.S.*
Department of Chemical Engineering, School of Engineering and Engineering Technology, Federal University of Technology, PMB 65 Minna, Niger State, Nigeria
Department of Civil and Chemical Engineering, College of Science, Engineering and Technology, University of South Africa, Private Bag X6, Florida 1710, Johannesburg, South Africa

Afolabi A.S.
Department of Civil and Chemical Engineering, College of Science, Engineering and Technology, University of South Africa, Private Bag X6, Florida 1710, Johannesburg. South Africa

Acknowledgment

Step B project, Federal University of Technology, Minna Nigeria is appreciated for their support. National Research Foundation (NRF), South Africa and Faculty of Science, Engineering and Technology, University of South Africa are also appreciated for their support.

5. References

[1] Abdulkareem, A.S.; Odigure, J.O. & Kuranga. M.B. (2010). Production and Characterization of Bio-Fuel from Coconut oil. Energy Source Part A. J. 32 106-114.
[2] Sanchez, O.J. and Cardona, C.A. (2008). Trends in Biotechnical Production of Ethanol Fuel from Different Feedstocks, *Bioresour. Technol.*, 37(2): pp. 133–140
[3] Balat, M. and Balat, H. (2009). Recent Trends in Global Production and Utilization of Bioethanol Fuel, *Applied Energy Journal*, 86: pp. 2273-2282.
[4] U.S. Department of Energy: Office of Fossil Energy. Retrieved from http://www.energy.gov/energysources/fossilfuels.htm/ on April 22, 2011.
[5] Alamu, O.J. (2007). Effect of Ethanol-Palm Kernel Oil Ratio on Alkali-Catalyzed Biodiesel Yield, Pacific Journal of Science and Technology, 8(2), pp 212-219.

* Corresponding Author

[6] Munack, A., Krahl, J., Baum, K., Hackbarth, U., Jeberien, H.E., Schutt, C., Schroder, O., Walter, N., Bunger, J., Muller, M.M. and Weigel, A. (2001). Gaseous Compounds, Ozone Precursors, Particle Number and Particle Size Distributions, and Mutagenic Effects Due to Biodiesel, *Trans. ASAE* 44(3): pp. 179–191.

[7] Knothe, G. (2002). Analyzing Biodiesel: Standards and other Methods, *Journal of the American Oil Chemist's Society*, 83(10): pp. 823–833.

[8] Graboski, M.S., McCormick, R.L. (1998). Combustion of fat and vegetable oil derived fuels in diesel engines. Progress in Energy Combustion Science. J. 24, 125–164.

[9] Berchamns H.J and Hirata S (2008). Biodiesel Production from Crude Jatropha Curcas L Seed Oil with a high content of free fatty acids. Bioresources Technology. J. 99 1716-1721.

[10] Adeniyi, O.D.; Kovo, A.S.; Abdulkareem. A.S & Chukwudozie. C. (2007): Ethanol Fuel Production from Cassava as a Substitute for Gasoline. Dispersion and Technology.J. 28 501-504.

[11] Carraretto, C.; Macor, A. & Mirandola, A. (2004). Biodiesel as alternative fuel: Experimental analysis and energetic evaluation. Energy. J. 2195-2211.

[12] Gerpen, V.J. (2005). Biodiesel Processing and Production. Fuel Processing Technology. J. 86(10) 1097–1107.

[13] Helwani, Z.; Othman, M. R,; Aziz, N.; Fernando, W. J. N.& Kim. J.(2009). Technologies for production of biodiesel focusing on green catalytic techniques: A review. Fuel Processing Technology.J. 90 1502-1514.

[14] Aghan D. (2005). Biodiesel production from vegetable oils via catalytic and non catalytic supercritical methanol tranestarification methods. rogress in energy and combustion. J. 31 406-487.

[15] Abdulkareem A.S., Uthman H., Afolabi A.S and Awonebe O.L (2011). Extraction and Optimization of Oil from Moringa Oleifera Seed as an Alternative Feedstock for the production of Biodeisel. Majid N, Mostafa K, editors. Sustainable Growth and Application in Renewable Energy Sources. InTech. Pp243-268.

[16] Abigor, R.D., Uadia, P.O., Foglia, T.A. Haas, M.J., Okpefa, J.E. and Obibuzor, J.U. (2000). Lipase-Catalyzed Production of Biodiesel Fuel from Nigerian Lauric Oils, *Biochem. Soc. Trans.* 28(3): 979–981.

[17] Eevera, T.; Balamurughan, P.; and Chittibabu .S. (2011). Characterization of groudnut oil based biodiesel to assess the feasibility for power generation. Energy Sources, Part A. J. 33 1354-1364.

[18] Okoro N. Linus, Sedoo V. Belaboh, Nwamaka R. Edoye, Bello Y. Makama (2011): Synthesis, Calorimetric and Viscometric Study of Groundnut oil Biodiesel and Blends. Research Journal of Chemical Sciences, 1(3).

[19] Yusuf N. and Sirajo, M. (2009). An Experimental Study of Biodiesel Synthesis from Groundnut Oil, *Aus. J. Applied Sci.*, 3: pp. 1623–1629.

[20] Saravanan, S., Nagarajan, G., Rao, G. L. N. and Sampath, S. (2007): Feasibility study of crude rice bran oil as a diesel substitute in a DI-CI engine without modifications, Energy for Sustainable Development, 11(3): 83- 95.

[21] Biomass Research and Development Board (2008). The economics of biomass feedstocks in the United States: A review of the literature. Retrieved from

http://www.biodiesel.org/ on October 12, 2008.

[22] Ahmmad, M.; Ullah, K.; Khan, M.A.; Zafari, M.; Tariq, M.; Ali, S. & Sultana. S. (2011). Physico chemical analysis of hemp biodiesel: A Promising non edible new sources for bioenergy. Energy Sources, Part A. J. 33 1365-1374.

[23] Geller, D.P., Goodrum, J.W. and Campbell, C.C. (1999). Rapid Screening of Biologically Modified Vegetable Oils for Fuel Performance, *Trans. ASAE*, 42(4) 859–862.

[24] Chavanne, C.G. (1938). Belgian Patent 422,877, Aug. 31, 1937; *Chem. Abs. 32*:4313.

[25] Freedman, B., Butterfield, R.O. and Pryde, E.H. (1986). Transesterification Kinetics of Soybean oil, *Journal of the American Oil Chemists Society*, 63(10): pp. 1375–1380.

[26] Diesel, R. The Diesel Oil-Engine and Its Industrial Importance Particularly for Great Britain, *Proc. Inst. Mech. Eng.* pp. 179–280 (1912); *Chem. Abstr.* 7(1): pp. 1605 (1913).

[27] Canakci, M. and Gerpen, V.J. (2001). Biodiesel Production from Oils and Fats with High Free Fatty Acids, *Transactions of the ASAE*, 44(6): pp. 1429–1436.

[28] Shay, E.G. (1993). Diesel Fuel from Vegetable Oil: Status and Opportunities, *Biomass Bioenergy*, 4(4): pp. 227-242.

[29] Lang, X., Dalai, A.K., Bakhshi, N.N., Reaney, M.J. and Hertz, P.B. (2001). Preparation and Characterization of Bio-diesels from Various Bio-oils, *Bioresource Technology*, 80(4): pp. 53–62.

[30] Okamura, K. Substitute Fuels for High-Speed Diesel Engines, *Journal. Fuel Soc. Japan*, 19(1): pp. 691–705 (1940); *Chem. Abstract* 35: 19647 (1941).

[31] Ma, F., Clements, L.D. and Hanna, M.A. (1998). The Effects of Catalyst, Fatty Acids and Water on Transesterification of Beef Tallow, *Transactions of the ASAE*, 41(4): 1261-1264.

[32] Sambo, A.S. (2007). Renewable Energy Development in Nigeria: A Situation Report. *In*: Proceedings of the International Workshop on Renewable Energy Development in Africa, July 30th to August 1st, University of Nigeria, Nsukka, 1–39.

[33] Ford, G.H. (1921). Vegetable Oils as Engine Fuel, *Cotton Oil Press* 5: pp. 38; *Chem. Abstract* 15: 3383.

[34] Korbitz W., (1999). Biodiesel Production in Europe and North America, an Encouraging Prospect, *Renew Energy* 16(2): 80–83.

[35] Lumet, G. and Marcelet, H. (1927). Utilization of Marine Animal and Fish Oils (as Fuels) in Motors, *Compt. Rend.* 185: pp. 418–420; *Chem. Abstr.* 21: 27–37.

[36] Sanchez, O.J. and Cardona, C.A. (2008). Trends in Biotechnical Production of Ethanol Fuel from Different Feedstocks, *Bioresour. Technol.*, 37(2): 133–140

[37] Dalai A.K & Kulkarni, M.G. (2006). Waste cooking oil-an economical source for biodiesel: A review. Industrial & Engineering Chemistry Research. J. 45 2901-2913.

[38] Ahmad, M.; Samuel, S.; Zafari, M.; Khan, M.A.; Tariq. M.; Ali. S. & Sultana. S. (2011). Physicochemical characterization of eco-friendly rice bran oil biodiesel. Energy, Part A. J. 33 1386-1397.

[39] Durosoy, I.; Turker, M. F.; Keles, S. & Kaygusuz, K. (2011). Sustainable agriculture and the production of biomass for energy use. Journal of Energy Sources, Part A, 33, 938-947.

[40] Hamamci, C.; Saydut, A.; Tonbul, Y.; Kaya, C. & Kafadari, A.B (2011). Biodiesel production via transestarification from safflower (Carthemustinctorious L) seed oil. Energy Sources Part A. J. 33 512-520.

[41] Pearl, G.G. (2002). Animal Fat Potential for Bioenergy Use, *Bioenergy*, The Tenth Biennial Bioenergy Conference, Boise, ID, Sept. 22-26.

[42] Evans, C.D., Moser, H.A., Cooney, P.C., Cowan, J.C. (1966). The Stability of Soybean Oil: Effect of Time and Temperature on Deodorization, *JAOCS*, 43(2): 632-634.

[43] U.S. Department of Agriculture (2009). Data on Groudnut oil. Retrieved from http://www.nal.usda.gov/fnic/foodcomp/search/ on April 22, 2011.

[44] Abulu, G.O. (1978). An Economic Analysis of Groundnut Production in Northern Nigeria. *In*: Proceedings of the National Seminar on Groundnut Production, Kano, Nigeria. pp. 83-90.

[45] Hill, P.D., (2002). Biodiesel Basics. Retrieved from http://www.biodieselgear.com/ on August 16th, 2002. 7-10.

[46] AOCS (1998). Official Methods and Recommended Practices of the American Oil Chemists Society. Champaign, IL

[47] Ibeto, C.N., Ofoefule, A.U. and Ezeugwu, H.C. (2011). Fuel Quality Assessment of Biodiesel Produced from Groundnut Oil (Arachis hypogea) and its Blend with Petroleum diesel, *American Journal of Food Technology*, 6(9): 798-803.

[48] Al-zahrani, K.M. (2005). Determination of saponification number in olive oil. 14(3) pp13-27

[49] Othmer, K. (2011). Castor Oil: Encyclopedia of Chemical Technology. Retrieved from http://www.castoroil.in on 25th June, 2011.

[50] Pasqualino, J. C., Montane, D. and Salvado, J. (2006). Synergic Effects of Biodiesel in the Biodegradability of Fossil-Derived Fuels, *Biomass Bioenergy*, 30(2): 874–879.

[51] Anitha, A. and Dawn, S.S. (2010). Performance characteristics of biodiesel produced from waste groundnut oil using supported heteropolyacids, *International Journal of Chemical Engineering and Applications*, Vol.1, No. 3, 261-265.

[52] Freedman, B., Butterfield, R.O. and Pryde, E.H. (1986). Transesterification Kinetics of Soybean oil, *Journal of the American Oil Chemists Society*, 63(10): 1375–1380.

[53] Galadima, A., Garba. Z.N. and Ibrahim, B.M. (2008). Homogeneous and Heterogeneous Transesterification of Groundnut Oil for Synthesizing Methyl Biodiesel, *International Journal of Pure and Applied Sciences*, 2(3): 138-144.

[54] Graboski, M.S. and McCormick, R.L. (1998). Combustion of Fat and Vegetable Oil Derived Fuels in Diesel Engines, *Prog. Energy Combust Sci.*, 24(3): 125–164.

[55] Gunstone, F.D. (2004). Rapeseed and Canola Oil: Production, Processing, Properties and Uses, *JAOCS*, 33: 132–139.

[56] Pintoa, A. C., Guarieiroa, L. L. N., Rezendea, M. J. C., Ribeiroa, N. M., Torresb, E. A., Lopesc, W. A., Pereirac, P. A. P. and Andrade, J.B.(2005): Biodiesel: An overview, J. Braz. Chem. Soc., Vol. 16, No. 6B, 1313-1330.

[57] Dorado, M.P., Ballesteros, E., Lopez, F.J. and Mittelbach, M. (2004). Optimization of Alkali-Catalyzed Transesterification of *Brassica carinata* Oil for Biodiesel Production, *Energy Fuel*, 18(2): 77–83.

Chapter 9

Optimalization of Extraction Conditions for Increasing Microalgal Lipid Yield by Using Accelerated Solvent Extraction Method (ASE) Based on the Orthogonal Array Design

Lin Rulong, Cai Wenxuan, Xing Bingpeng and Ke Xiurong

Additional information is available at the end of the chapter

http://dx.doi.org/10.5772/52475

1. Introduction

Since the fossil fuel crisis broke out in the nineteen seventies with the continual rise in fossil fuel prices, the mankind has been searching for renewable energy for consumption. For the past decades, atmospheric pollutions that involve with using fossil fuel have resulted in many severe problems of environment and human health[1-16]. Therefore, exploitation and utilization of clean and renewable energy have become the strategic consideration for many countries. Biofuel, as an alternative fuel, is recently attracting increasing attention[17-20]. Microalgae grow in aquatic environments and use light and carbon dioxide to create biomass and have been recognized as potentially good material sources for biofuel production. Microalgae possess several aspects of advantages for development of clean green energy, *i.e.* they have short growth period and are easy to cultivate and reproduce to large biomass. Controlled culture conditions of microalgae could trigger high lipid content of microalgae which could be used for lipid extraction, in turn, by further transesterification reaction, for preparation and production of biofuel with excellent characteristics. Therefore, utilization of the microalgal lipid for producing biofuel has promising future[21-27]. It should be noticed that obtaining lipid is a prerequisite for the production of microalgal fuel. More and more investigations have showed that microalgae are potentially good biomass materials for development of clean green energy[28-40].

As shown in figure 1, there are six major steps for biofuel preparation and production from initiating microalgal cultivation to biofuel products. Each of six steps involves with crucial techniques and methods in order to achieve high production of biofuel. For example, during microalgal cultivation, it is very important to screening and selecting fine microalgal strains with higher oil content for reproduction and amplification of algal cells[41-45]. In addition, investigating on controlled culture conditions that could improve oil accumulation of microalgal cells is necessary for the increase in biofuel production[46-51]. Usually, open ponds are used in microalgal cultivation with the less expense, simple facilities and operation. However, its disadvantages are associated with such issues as culture contamination, difficulties in regulation and control of culture conditions(like temperature and light control), lower productivity and so on[52-54]. Closed system such as photobioreactors(PBRS) are as well used for microalgal culture. These are highly-automated clear piping systems, which allow the operator to control nutrients, light, temperature and contamination for high productivity. But such facilities require expensive investment[55-59]. While heterotrophic culture and amplification of algal cells by using fermentation tanks can obtain highly-concentrated algal cells for high productivity effects, such facilities are also involved in high investment cost[60-61]. Microalgal harvesting is another crucial technique for entire biofuel production process. Since microalgal cells are so tiny (only micron order of magnitude in size) that it is quite difficult for effective microalgal cell harvesting. Taking cost and energy efficiencies into consideration, a relative simple and feasible method, flocculation of microalgal cells by changing pH value of culture medium or using certain eco-friendly chemical and biological flocculants like ferric chloride and chitosan, could be adapted in harvesting microalgal biomass. During flocculation, the dispersed microalgal cells can aggregate and form larger conjugates with higher sedimentation rate. Moreover, those methods allows the cycle reuse of the flocculated medium, thereby contributing to the economic cultivation and harvest of microalgae[62-66]. Harvested microalgal cells can further be made in form of powers through the process of dehydrating and drying. Optimized treatment conditions regarding extraction and transesterfication reaction of microalgal oils, which need to be further developed and explored, as well play an important role in the effective biofuel target products[67-70].

Due to miniature and hardiness of microalgal cells, it is usually difficult for the extraction of microalgal lipid component which often requires the operation of special treatment (such as cell wall breaking, pressurizing and heating etc.) to achieve more complete extraction effect. Traditionally, there could be several methods used in extraction of microalgal lipid to get information on lipid content of biological samples, for instance, the Soxhlet extraction method by using organic solvents for biological sample treatment and sample-heating treatment with some strong inorganic acids and so on. [71-74].

In spite of simplicity and easiness regarding those methods, obvious disadvantages are time-consuming for the analysis and operational treatment. Moreover, a considerable amount of organic solvents or other acid substances, which could involve with human health problem and the pollution of the environment, are often used in a sample-treating process. Supercritical CO2 extraction method for extracting biological sample lipid is quite effective, but it requires expensive equipment to complete sample analysis[75-76]..

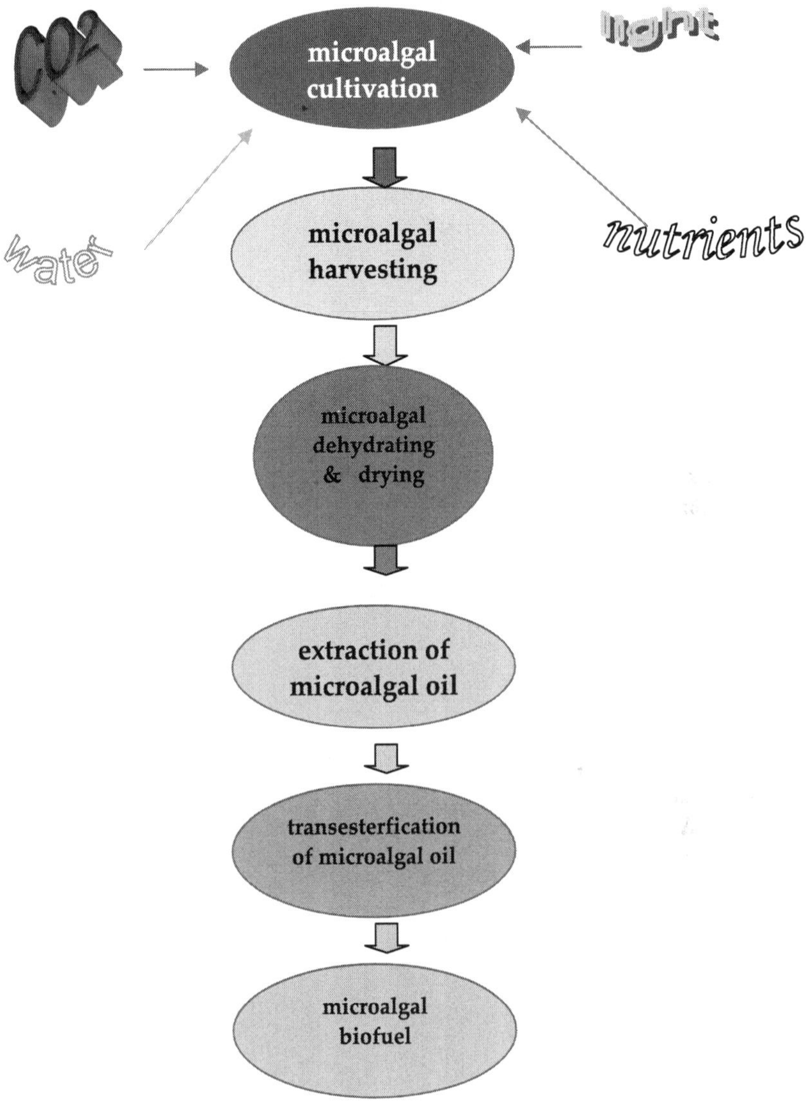

Figure 1. Schematic diagram of biotechnical process of microalgal biofuel preparation

In order to consume less toxic, less amount of organic solvents and obtain higher algal lipid yield result, it is crucial to adopt appropriate methods and conditions for lipid extraction of microalgal materials. Accelerated solvent extraction (ASE) is one of the best methods for extraction of microalgal lipid with a small amount of organic solvents needed in extraction treatment. In this investigation, we carried out the experimental study on optimalization of extraction conditions for increasing microalgal lipid yield by using accelerated solvent extraction method based on applying an orthogonal array design (OAD). Experimental factors including extraction solvent (hexane, chloroform, petroleum ether, ethanol, acetone), temperature (75-175 °C), time (4–20 min) and extraction cycle number (1–5) at five-levels were studied in 25 trials by OAD_{25} (5^6) to reach rapid and high lipid extraction for the marine microalga （*Nannochloropsis oculata* Droop）.

The objectives of this study were:

1) to determine which factors might have more significant effects than the others on the extraction of microalgal lipid;2) to obtain the optimum level of each tested factor; and 3) to determine a best combination of the 4 tested factors with 5 factoral levels to be used as increasing extraction efficiency for microalgal lipid yield.

2. Materials and methods

2.1. Cultivation and treatment of microalgal species for experiment study

The strain of marine microalga（*Nannochloropsis oculata* Droop）was from our laboratory storage and used for batch culture step by step to sufficient quantity of algal cells. The microalga was cultivated with general enriched seawater f/2-Si medium designed for growing coastal marine algae (Guillard and Ryther 1962). The microalga was grown under regulated and controlled conditions(water temper 25C, light intensity 5000lux, salinity 30‰,PH 7.8) and harvested during log growth phase. Microalgal cells from collection liquid were condensed by a centrifugal treatment process and desalinated after two times of distilled water washing and centrifugal treatment and prepared in form of microalgal powder by using freeze drying process for extraction of microalgal lipid.

2.2. Chemical reagents and intrument used in the experiment

All chemicals and reagents used in this experimental study were analytical or research grade without further purification and from Xiamen Luyin Chemical Company. Intrument accelerated solvent extractor ASE 100 (Dionex) was used for extraction of microalgal lipid.

2.3. Experimental designation of ASE method for the extraction of microalgal lipid

Four factors with five levels each were designed for their effects to be investigated on the extraction of microalgal lipid with orthogonal array design. An orthogonal array table OAD_{25} (5^6) was used for designing ASE experiment of microalgal lipid extraction. Experimental designation for different factors and levels influencing the extraction of microalgal lipid was arranged in following table 1.

level \ factor	A extraction solvent	B extraction tem. ◎	C extraction time min	D extraction cycle times
1	hexane	75	4	1
2	chloroform	100	8	2
3	petroleum ether	125	12	3
4	ethanol	150	16	4
5	acetone	175	20	5

Table 1. Designation of factors and levels for lipid extraction by ASE method

2.4. Operational method of ASE extraction and instrument analytical conditions

Extraction operation process of microalgal lipid and parameter settings: Appropriate amount of about 5g (5.120 ± 0.076g) of microalgae powder samples was put into 34ml extraction pool of the instrument. The extraction pressure value was constant at 1500psi. Based on combination of different factors and levels of five types of different extraction solvents (hexane, chloroform, petroleum ether, ethanol, acetone), extraction temperature range from 75~175 °C, extraction time range from 4~20 min, extraction cycle number for 1~5 times, corresponding operational treatment of ASE microalgal lipid extraction was adopted according to 4 factors and 5 levels of orthogonal experiment set(refer to Table2). Other relevant extraction parameters were constantly set as 60% of flush volume and 90s of purge time for microalgal lipid extraction. Microalgal lipid extracted was steam-dried by a rotary evaporator and further dried via N2 gas blowing process and finally dried at 100 °C for 2 hours. The final microalgal lipid quantity extracted for different ASE operation was expressed as lipid % based on algal dry weight.

2.5. Conventional Soxhlet extraction method for microalgal lipid

The extraction of microalgal lipid was concurrently conducted by conventional classic Soxhlet and using same extraction solvents to compare extraction efficiency with ASE method. Appropriate ammout of microalgal power samples mixed with quartz sand particles was ground in a mortar and then transferred to extraction cylinder of the extractor. Solvent extraction included the process with 18 hours of Static extraction and 6 hours of dynamic extraction to reach a thorough extraction. The final microalgal lipid quantity extracted was expressed as lipid % based on algal dry weight.

2.6. Calculation of extraction efficiency increase based on ASE and Soxhlet methods for microalgal lipid

Extraction efficiency increase(EI%) was calculated by formula below:

$$EI(\%)=100\times A\text{-}S)/A$$

A and S respectively represent the lipid amount (gram) of microalga extracted by the methods of ASE and Soxhlet.

2.7. Data analysis and treatment

The arrangement of importance of the four factors to the extraction of microalgal lipid were evaluated according to the effectiveness of each factor through the calculation of ranges (R value) (determined from the difference between the maximal and minimal lipid content (%) within the five levels of each factor), that means, the factor with the most effectiveness (i.e., with the largest range of R value) to the extraction of microalgal lipid is considered as the most important factor, the factor with the lest effectiveness (i.e., with the smallest range of R value) to the extraction of microalgal lipid is considered as the lest important factor. Analysis of variance (ANOVA) was conducted to test the significance of the effects of the four factors on the extraction of microalgal lipid by using statistical software SPSS 15.0. In all analyses, the level of significance was set at a P-value of 0.05.

3. Results

3.1. The orthogonal experiment result and analysis

The orthogonal experiment result and analysis based on 4 factors and 5 levels was shown in Table 2 for ASE extraction of microalgal lipid and the associated variance analysis result shown in Table 3. Variation trend of extraction efficiency of microalgal lipid was shown in Figure 2 for different extraction operations with various factor level values.

The experimental results indicated that: by using different extraction solvents and various combinations of different extraction operations, lipid content (%) had the apparent difference (range between 2.98%~21.36%). This suggests that different extraction treatments on microalgal cells result in the difference in lipid yield of the microalga. Solvents chloroform, hexane and petroleum ether had normally poor extraction effect on microalgal cells, and the anhydrous ethanol and acetone were good extraction solvents for microalgal lipid. Calculation results of range of R value based on table 2 test experiment reflect the size of the corresponding factor effect. Compared to those factors with smaller R value, the factors with greater R value are generally significant factors to make remarkable influence on lipid extraction of microalgal cells since more difference of lipid yield occurs at the different levels of those factors. Our experimental results showed that, the R values caused by the extraction solvent, extraction temperature, extraction time and extraction cycles were respectively 25.91, 38.85, 16.44 and 16.67. Therefore, according to the size of the R values, the significance of test factors for accelerated solvent extraction (ASE) of microalgal lipid may be arranged as: extract temperature, extraction solvent, extraction cycle, extraction time.

Variance analysis of the results of accelerated solvent extraction (ASE) processing experiment data further indicated (Table 3), extraction effect of temperature on the microalgal lipid was significant (P =0.000515), followed by significant extraction effect of solvents (P =0.003855).The significant extraction effect of extraction time at significant level of a =0.05 was also observed (P =0.035094). Comparatively, the effect of extraction cycles on microalgal lipid was relatively small and it was not significant (P =0.081996) at the significance level set for a =0.05.

factor / trial no	A extraction solvent	B extraction T ◎	C extraction time min	D cycle times	lipid (%) (alga DW)
1	hexane	75	4	1	2.98
2	hexane	100	8	2	5.76
3	hexane	125	12	3	8.52
4	hexane	150	16	4	15.74
5	hexane	175	20	5	16.15
6	chloroform	75	8	3	10.05
7	chloroform	100	12	4	10.21
8	chloroform	125	16	5	13.37
9	chloroform	150	20	1	13.10
10	chloroform	175	4	2	13.92
11	petroleum ether	75	12	5	5.81
12	petroleum ether	100	16	1	5.44
13	petroleum ether	125	20	2	13.32
14	petroleum ether	150	4	3	10.77
15	petroleum ether	175	8	4	13.45
16	ethanol	75	16	2	12.25
17	ethanol	100	20	3	16.09
18	ethanol	125	4	4	14.87
19	ethanol	150	8	5	15.99
20	ethanol	175	12	1	15.86
21	acetone	75	20	4	10.79
22	acetone	100	4	5	10.47
23	acetone	125	8	1	12.74
24	acetone	150	12	2	13.07
25	acetone	175	16	3	21.36
level I	49.15	41.89	53.01	50.12	∑=302.09
level II	60.65	47.97	58.00	58.33	
level III	48.80	62.82	53.47	66.79	
level VI	75.06	68.68	68.17	65.06	
level V	68.43	80.74	69.45	61.80	
Value R	25.91	38.85	16.44	16.67	

Table 2. Result and analysis of orthogonal experimentation by ASE method

source	SS	df	MS	F	p	significance
A. solvent	108.117	4	27.029	9.564	0.003855	***
B. temperature	197.100	4	49.275	17.435	0.000515	***
C. time	50.116	4	12.529	4.433	0.035094	**
D. cycle	34.878	4	8.719	3.085	0.081996	
error	22.610	8	2.826			
sum	412.820	24				

SS =Sum of squares; df = degrees of freedom; MS =mean squares

Table 3. Variance analysis of orthogonal experimentation by ASE method

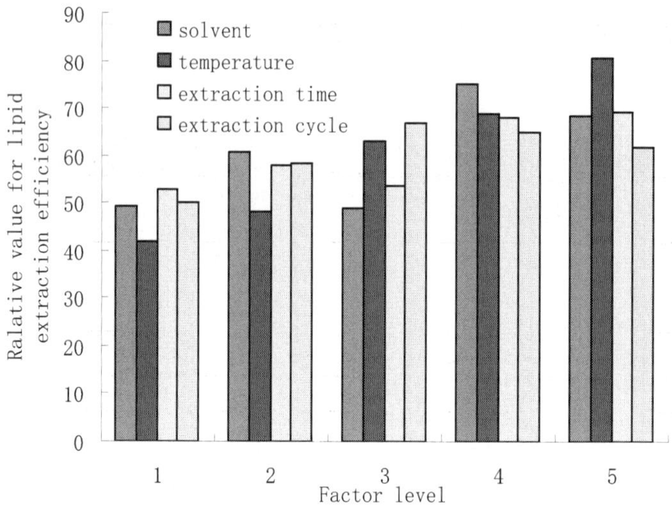

Figure 2. Variation trend of lipid extraction for different factors and levels

Figure 2 also demonstrated the variation trend of extraction effect of microalgal lipid for different factors and levels of operational conditions based on ASE method. For extraction solvents, ethanol and acetone had the best extraction effect for microalgal lipid extraction, followed by hexane and chloroform and solvent petroleum had the poorest extraction effect for microalgal lipid. The lipid yield raised with an increase in extraction temperature or extraction time and reached the maximum at a temperature of 175 ℃,extraction time of 16min and 3 extraction cycles. Therefore for operational simplicity, it was not necessary for extraction process of microalgal lipid to take more than 16min and 3 extraction cycles.

Taking into consideration the optimal extraction effect for lipid yield of microalgal cells based on the R values of orthogonal experimental data and the results of variance analysis of factors and levels, the best operational parameters for ASE method are: using extraction solvents of ethanol or acetone, extraction temperature of 175◎, extraction time of 16min and 3 extraction cycles, which resulted in the highest lipid production. For the health and cost consideration, it is more preferable for using ethanol in lipid extraction operation in due to its relatively less toxicity and price.

3.2. Validation of optimal ASE extraction conditions and comparison of extraction effectiveness

Based on the results of orthogonal experiment and data analysis, it was observed that optimized ASE treatment for the extraction of microalgal lipid was using ethanol or acetone as solvents with other operational parameters such as 1500psi of extraction pressure, 175℃ of extraction temperature, 16 minutes of extraction time and three extraction cycle. To evaluate the stability and superiority of optimized ASE treatment effect for the extraction of microalgal lipid, a comparison was made between the optimized ASE treatment and conventional Soxhlet method for the extraction effectiveness of microalgal lipid. The results were shown in Table 4.

Trial #	ethanol (ASE)				acetone (ASE)			
	Microalgal powder (g.dw)	Extracted lipid(g)	Lipid %based on micro-algal dw	AI%	Microalgal powder (g.dw)	Extracted lipid(g)	Lipid %based on microalgal dw	AI%
1	5.0578	0.9805	19.57	44.19	5.0140	0.9810	19.57	39.08
2	5.0637	1.0356	20.45	47.09	5.0616	1.0694	21.13	43.58
3	5.0504	0.9908	19.62	44.85	5.0521	1.0012	19.82	39.85
4	5.0838	0.9779	19.24	43.75	5.2892	1.0763	20.35	41.42
5	5.0735	0.9739	19.20	43.63	5.0856	1.0658	20.96	43.12
6	/	/	/	/	5.0040	1.0687	21.36	44.19
($\overline{X}$)	5.0658	0.9917	19.62	44.70	5.0844	1.0437	20.53	41.87
(±SD)	0.0131	0.0253	0.50	1.42	0.1048	0.0414	0.73	2.10

Note: For conventional Soxhlet method, 1.0 gram of microalgal powder was used for lipid extraction using same extraction solvents. Microalgal yields were respectively 0.1082 gram with calculated lipid % as 10.82% (based on algal dry weight) for using ethanol extraction and 0.1192 gram with calculated lipid % as 11.92% (based on algal dry weight) for using acetone extraction.

Table 4. Validation of optimal ASE extraction conditions and comparison of extraction effectiveness

Table four results clearly showed that the fluctuation of lipid extraction yield was very small and extraction effect was quite stable for optimum processing conditions of accelerated solvent extraction (ASE) for microalgal lipid extraction. Not only were the extraction process time and extraction solvent volume considerably saved, but also lipid extraction effectiveness were greatly improved in ASE method. Compared with conventional Soxhlet extraction method, ASE method with ethanol as extraction solvent, extraction efficiency could increase 43.63-47.09% (mean ± SD 44.70 ± 1.42%). For using acetone as the extraction solvent extraction efficiency of ASE method could increase 39.08-44.19% (mean ± SD to 41.87 ± 2.10%). Therefore, adopting the optimum processing conditions of ASE method for microalgal lipid extraction can reach maximum microalgal lipid yield and its lipid extraction efficiency is obviously higher than conventional Soxhlet extraction method.

4. Discussion

The orthogonal array design is a useful experiment methodology,especially for multi-factor experiment and analysis. It can provide useful and sufficient information for accessing and evaluating main factors and the optimum combination of factor levels for target parameter as less experimental trials as possible [77-80]. ASE method is approved for use by the U.S. EPA and CLP Program and a good one for extracts in treating many different samples. Extractions that normally take hours can be done in minutes using Accelerated Solvent Extraction (ASE). Compared to techniques like Soxhlet and sonication, ASE generates results in a fraction of the time. In addition to speed, ASE offers a lower cost per sample than other techniques by reducing solvent consumption by up to 90%. Relatively less extraction time, reduction in solvent consumption and wide range of application are the essencial advantages of ASE method. By using conventional liquid solvents at elevated temperatures and pressures, ASE increases the efficiency of the extraction process. Increased temperature accelerates the extraction kinetics, and elevated pressure keeps the solvent below its boiling point, thus enabling safe and rapid extractions. Although ASE uses the same aqueous and organic solvents as traditional extraction methods, it uses them more efficiently. ASE can be used to replace Soxhlet, sonication, wrist shaking, and other extraction techniques typically used[81].

Because the ASE method requires relatively simple equipment with many aspects of advantages such as higher degree of automation, good safety, less solvent consumption, fast-complete extraction and high efficiency , it has been widely applied in analyzing and testing various types of samples from different sources. For instance, it can be used to detect the extracts from the water, soil, sediment, minerals, chemical products, biological samples (vegetables, fruits, meat, fish, plants) and other harmful substances (such as various pesticides, hydrocarbons, chemicals and the like) [82-93]. ASE technology is very important and helpful for environmental protection and human health. For determination and evaluation of bio-active components from animals and plants, especially for separation, extraction and purification of Chinese traditional herb medicines, ASE also played an important role [94-97]. For example, ASE method has been applied for extraction of phenolic acid compound salvia, volatile oil from Mu Xiang, almond oil from plants, saponin from Ginseng. Many relevant studies indicate that target product yield and extraction efficiency are higher by using ASE method than conventional types of extraction techniques[98-100]. Our present study also showed extraction efficiency for microalgal lipid has an increase of 39.08-47.09% by using ASE method, compared with conventional Soxhlet extraction method. This suggest that ASE technology has the wide applicability in different fields of sample extraction.

Due to different features and characteristics with extract of the target products, the application of ASE method should depend on the actual situations and determine appropriate parameter settings for extraction processing in order to obtain the practical optimal extraction results. For example, using methanol as solvent extraction with extraction parameters set as pressure 1500psi, temperature 140◎, time 5 minutes, extraction cycle number 2, ginseng saponins could reach the maximum extraction amount, which was 25.88-58.68% of higher than other conventional extraction methods (such as immersion method, ultrasonic method, homogenization, mechanical vibration method) [98]. Zhang *et al* (2007) reported the results of extraction of flavonoid compounds in citrus peels using optimum ASE operational conditions and showed that maximum extraction rate of target products was obtained with using 80% ethanol as solvent, pressure 10.3Mpa, temperature 70 ◎, time of 10 minutes, extraction cycle number 1 [99] . Pang *et al* (2007) used uniform experimental design method and reported extraction of almond kernel oil with the optimum ASE process operations. The results showed that the maximum amount of oil extraction was obtained by using acetone: hexane (1:3) as solvent, the temperature of 120-140 ◎, time 6-12 minutes, extraction cycle number 1-3[100]. Herrero (2005) studied the extraction of bioactive products for microalga (*Spirulina platensis*).The optimal ASE extraction conditions of antioxidant compounds of the microalga were using ethanol as solvent, temperature 170◎, time 3-9 minutes, which resulted in target product extraction as high as 19.7% (based on algal dry weight) compared to only 2.94-8.22% of antioxidant compound extraction by using the other three types of solvents (hexane, petroleum ether, water) in the same extraction conditions [101]. This suggests that extraction parameter setting for target product extraction is crucial for the extraction result of applying ASE method. In summary, determination of the optimal extraction conditions should depend on different samples, target products and the experimental designs in application of ASE method.

Our present study was involved with the extraction of microalgal lipid by using ASE method. This study clearly showed that an increase of temperature and pressure during the extraction process could greatly enhance the solvent penetration and diffusion capacity, thereby result in a rapid extraction of microalgal lipid components. Compared with the conventional Soxhlet extraction method, ASE method with the optimal operational conditions could significantly improve the microalgal lipid extraction and raise 39.08-47.09% of lipid extraction efficiency. For a consideration of security and practicability, using ethanol or acetone with the lowest toxicity as extraction solvents is another advantage for sample treatment. Therefore, using ASE method with the optimization of extraction conditions is suitable for the rapid and efficient extraction of microalgal lipid.

5. Conclusion

Our findings in the present investigation demonstrated that accelerated solvent extraction method (ASE) based on the orthogonal array design is an effective approach for the extraction and determination of lipid content in biological microalgal samples. This study also demonstrated that the application of multiple-factor and level experimental design based on Taguchi's orthogonal array could determine the optimal extraction operation and obtain maximal yield for lipid extraction of microalgae.

Author details

Lin Rulong, Cai Wenxuan, Xing Bingpeng and Ke Xiurong
Key Laboratory of Global Change and Marine-Atmospheric Chemistry,
Third Institute of Oceanography,State Oceanic Administration, Xiamen, China

Acknowledgement

This study was supported by special research project fundings of China National Marine Public Welfare Industry (grant number200705025 and grant number 200705025).

6. References

[1] Abdus Salam, Hassan Al Mamoon, Md. Basir Ullah, Shah M. Ullah. Measurement of the atmospheric aerosol particle size distribution in a highly polluted mega-city in Southeast Asia (Dhaka-Bangladesh). *Atmospheric Environment.* 2012, 59:338-343.

[2] Jing Liu, Xiaoqian Ma. The analysis on energy and environmental impacts of microalgae-based fuel methanol in China. *Energy Policy,* 2009, 37(4): 1479-1488.

[3] Kakali Mukhopadhyay, Osmo Forssell. An empirical investigation of air pollution from fossil fuel combustion and its impact on health in India during 1973–1974 to 1996–1997. *Ecological Economics,* 2005, 55(2): 235-250.

[4] Xiaoping Wang, Denise L. Mauzerall. Evaluating impacts of air pollution in China on public health: Implications for future air pollution and energy policies. *Atmospheric Environment,* 2006, 40(9): 1706-1721.

[5] Jasmin Honold, Reinhard Beyer, Tobia Lakes, Elke van der Meer. Multiple environmental burdens and neighborhood-related health of city residents. *Journal of Environmental Psychology,* 2012, 32(4): 305-317.

[6] Marilena Kampa,Elias Castanas. Human health effects of air pollution.*Environmental Pollution,* 2008, 151(2): 362-367.

[7] J. Schwartz. Long-Term Effects of Particulate Air Pollution on Human Health. *Encyclopedia of Environmental Health,* 2011, 520-527.

[8] G.D. Thurston. Outdoor Air Pollution: Sources, Atmospheric Transport, and Human Health Effects. *International Encyclopedia of Public Health,* 2008, 700-712.

[9] Kira Matus, Kyung-Min Nam, Noelle E. Selin, Lok N. Lamsal, John M. Reilly, Sergey Paltsev. Health damages from air pollution in China. *Global Environmental Change,* 2012, 22(1): 55-66.

[10] Kenneth Donaldson, William MacNee. Potential mechanisms of adverse pulmonary and cardiovascular effects of particulate air pollution (PM10). *International Journal of Hygiene and Environmental Health,* 2001, 203(5-6): 411-415.

[11] Jane V. Hall, Victor Brajer, Frederick W. Lurmann. Air pollution, health and economic benefits—Lessons from 20 years of analysis. *Ecological Economics,* 2010, 69(12): 2590-2597.

[12] Klara Slezakova, Dionísia Castro & Arlindo Begonha *et. al.* Air pollution from traffic emissions in Oporto, Portugal: Health and environmental implications. *Microchemical Journal*, 2011, 99(1): 51-59.

[13] B.R. Gurjar, A. Jain, A. Sharma, A. & Agarwal, *et. al.*Human health risks in megacities due to air pollution. *Atmospheric Environment*, 2010, 44(36): 4606-4613.

[14] Janet Currie, Matthew Neidell, Johannes F. Schmieder. Air pollution and infant health: Lessons from New Jersey. *Journal of Health Economics*, 2009, 28(3): 688-703.

[15] Leigh A. Beamish, Alvaro R. Osornio-Vargas, Eytan Wine. Air pollution: An environmental factor contributing to intestinal disease. *Journal of Crohn's and Colitis*, 2011, 5(4): 279-286.

[16] Ni Bai, Majid Khazaei, Stephan F. van Eeden, Ismail Laher. The pharmacology of particulate matter air pollution-induced cardiovascular dysfunction. *Pharmacology & Therapeutics*, 2007, 113(1): 16-29.

[17] Mata T M, Martins A A, Caetano N S. Microalgae for biodiesel production and other applications: a review. Renewable and Sustainable Energy Reviews, 2010, 14(1): 217-232.

[18] Hossain A B M, Salleh A. Biodiesel fuel production from algae as renewable energy. *American Journal of Biochemistry and Biotechnology*, 2008, 4(3): 250-254.

[19] Song D. H, Fu J J. and Shi D. J. Exploitation of oil-bearing microalgae for biodiesel. *Chinses Journal of Biotechnology*, 2008, 24(3): 341-348.

[20] Cantrell, K.B. and Walker, T.H. 2009. Influence of temperature on growth and peak oil biosynthesis in a carbon-limited medium by pythium irregular. *Journal of the American Oil Chemists Society* 86 (8):791–797.

[21] Chisti, Y. 2007. Biodiesel from microalgae. *Biotechnology Advances*. 25(3): 294–306.

[22] Cooney, M., Young, G. and Nagle, N. 2009. Extraction of bio-oils from microalgae.*Separation and Purification Reviews*. 38 (4): 291–325.

[23] He hongbo, Yao yisheng; Jiang laien. Research progress of biodiesel preparation. *Anhui Chemical Industry*, 2008, 34(6):7-10.

[24] Evan Stephens, Ian L. Ross, Jan H. Mussgnug, Liam D. Wagner, Michael A. Borowitzka, Clemens Chisti, Y. 2007. Biodiesel from microalgae. *Biotechnology Advances*, 25(3), 294-306.

[25] Gouveia, L., Oliveira, A.C. 2008. Microalgae as a raw material for biofuels production. *Journal of Industrial Microbiology & Biotechnology*, 36(2): 269-274.

[26] Posten, Olaf Kruse, Ben Hankamer. 2010. Future prospects of microalgal biofuel production systems. *Trends in Plant Science*. 15(10): 554-564.

[27] Schenk, P., Thomas-Hall, S., Stephens, E., Marx, U., Mussgnug, J., Posten, C., Kruse, O., Hankamer, B. 2008. Second Generation Biofuels: High-Efficiency Microalgae for Biodiesel Production. *BioEnergy Research*, 1(1): 20-43.

[28] Wang Y Y, Wang J N, Gu B J. Research progress of biodiesel preparation method. *Modernizing Agriculture*, 2011(3): 40-42.

[29] Robles M A, Gonzalez M P A, Est Eban C L, *et al.* Bio catalysis: Towards ever greener bio diesel production. *Biotechnology Advances*, 2009, 27 (4) : 398-408.

[30] Miao X L, Wu Q Y. Study on preparation of biodiesel from microalgal oil. *Acta Energiae Solaris Sinica*. 2007, 28(2): 219-222.

[31] Fajardo, A.R., Cerdan, L.E., Medina, A.R., Fernandez, F.G.A., Moreno, P.A.G.. and Grima, E.M., 2007. Lipid extraction from the microalga *Phaeodactylum tricornutum*. *European Journal of Lipid Science and Technology*. 109(2): 120–126.

[32] Hu, Q., Sommerfeld, M., Jarvis, E., Ghirardi, M., Posewitz, M., Seibert, M. and Darzins, A. 2008. Microalgal triacylglycerols as feedstocks for biofuel production:perspectives and advances. *Plant Journal*. 54 (4): 621–639.

[33] Lalman, J.A. and Bagley, D.M. 2004. Extracting long-chain fatty acids from a fermentation medium. *Journal of the American Oil Chemists Society*. 81 (2): 105–110.

[34] Lee, S.J., Yoon, B.D.and Oh, H.M., 1998. Rapid method for the determination of lipidfrom the green alga *Botryococcus braunii*. *Biotechnology Techniques*. 12 (7): 553–556.

[35] Liu, B.and Zhao, Z. 2007. Biodiesel production by direct methanolysis of oleaginousmicrobial biomass. *Journal of Chemical Technology and Biotechnology*. 82 (8):775–780.

[36] Liu, X.J., Jiang, Y.and Chen, F., 2005. Fatty acid profile of the edible filamentous cyanobacterium Nostoc flagelliforme at different temperatures and developmental stages in liquid suspension culture. *Process Biochemistry*. 40 (1): 371–377.

[37] Molina Grima, E., Robles Medina, A., Gimenez Gimenez, A., Sanchez Perez, J.A.,Garcia Camacho, F.and Garcia Sanchez, J.L., 1994. Comparison between extraction of lipids and fatty-acids from microalgal biomass. *Journal of the American Oil Chemists Society*. 71 (9): 955–959.

[38] Rittmann, B.E. 2008. Opportunities for renewable bioenergy using microorganisms.*Biotechnology and Bioengineering*. 100 (2): 203–212.

[39] Sheng, J., Vannela, R.and Rittmann,B.E. 2011. Evaluation of methods to extract and quantify lipids from *Synechocystis* PCC 6803. *Bioresource Technology*. 102: 1697–1703.

[40] Xu, H., Miao, X.L. and Wu, Q.Y. 2006. High quality biodiesel production from a microalga *Chlorella protothecoides* by heterotrophic growth in fermenters. *Journal of Biotechnology*. 126 (4): 499–507.

[41] Doan, T.T.Y., Balasubramanian Sivaloganathan B. and Obbard J.P. 2011. Screening of marine microalgae for biodiesel feedstock. *Biomass and Bioenergy*. 35: 2534-2544.

[42] Glacio S. Araujo, Leonardo J.B.L. Matos, Luciana R.B. Gonçalves, Fabiano A.N. Fernandes, Wladimir R.L. Farias. 2011. Bioprospecting for oil producing microalgal strains: Evaluation of oil and biomass production for ten microalgal strains. *Bioresource Technology*. 102(8): 5248-5250.

[43] Griffiths, M.J., Harrison, S.T.L. 2009. Lipid productivity as a key characteristic for choosing algal species for biodiesel production. *Journal of Applied Phycology*, 21(5), 493-507.

[44] T. Mutanda, D. Ramesh, S. Karthikeyan, S. Kumari, A. Anandraj, F. Bux. 2011. Bioprospecting for hyper-lipid producing microalgal strains for sustainable biofuel production. *Bioresource Technology*. 102(1): 57-70.

[45] Rodolfi L, Chini Zittelli G, Bassi N, Padovani G, Biondi N, Bonini G, Tredici MR.2009. Microalgae for oil: strain selection, induction of lipid synthesis and outdoor mass cultivation in a low-cost photobioreactor. *Biotechnol Bioeng*. 102(1):100-112.

[46] Guido Breuer, Packo P. Lamers, Dirk E. Martens, René B. Draaisma, René H. Wijffels. 2012. The impact of nitrogen starvation on the dynamics of triacylglycerol accumulation

in nine microalgae strains. *Bioresource Technology*. http://dx.doi.org/10.1016/j.biortech.2012.08.003.

[47] Lv, J.M., Cheng, L.H., Xu, X.H., Zhang, L., Chen, H.L. 2010. Enhanced lipid production of Chlorella vulgaris by adjustment of cultivation conditions. *Bioresource Technology*, 101(17): 6797-804.

[48] Pal, D., Khozin-Goldberg, I., Cohen, Z., Boussiba, S. 2011. The effect of light, salinity, and nitrogen availability on lipid production by Nannochloropsis sp. *Applied Microbiology and Biotechnology*, 90(4): 1429-41.

[49] Ramasamy Praveenkumar, Kalifulla Shameera, Gopalakrishnan Mahalakshmi, Mohammad Abdulkader Akbarsha, Nooruddin Thajuddin. 2012. Influence of nutrient deprivations on lipid accumulation in a dominant indigenous microalga Chlorella sp., BUM11008: Evaluation for biodiesel production. *Biomass and Bioenergy*. 37: 60-66.

[50] Santos, A.M., Janssen, M., Lamers, P.P., Evers, W.A., Wijffels, R.H. 2012. Growth of oil accumulating microalga *Neochloris oleoabundans* under alkaline-saline conditions. *Bioresource Technology*. 104: 593-599.

[51] Takagi, M., Karseno, Yoshida, T. 2006. Effect of salt concentration on intracellular accumulation of lipids and triacylglyceride in marine microalgae *Dunaliella* cells. *Journal of Bioscience and Bioengineering*, 101(3), 223-226.

[52] Nasrin Moazami, Alireza Ashori, Reza Ranjbar, Mehrnoush Tangestani, Roghieh Eghtesadi, Ali Orlando Jorquera, Asher Kiperstok, Emerson A. Sales, Marcelo Embiruçu, Maria L. Ghirardi. 2010. Comparative energy life-cycle analyses of microalgal biomass production in open ponds and photobioreactors. *Bioresource Technology*. 101(4): 1406-1413.

[53] Probir Das, Siti Sarah Aziz, Jeffrey Philip Obbard. 2011. Two phase microalgae growth in the open system for enhanced lipid productivity. *Renewable Energy*. 36(9): 2524-2528.

[54] Sheykhi Nejad. 2012. Large-scale biodiesel production using microalgae biomass of *Nannochloropsis*. *Biomass and Bioenergy*. 39: 449-453.

[55] Eleonora Sforza, Alberto Bertucco, Tomas Morosinotto, Giorgio M. Giacometti. Photobioreactors for microalgal growth and oil production with *Nannochloropsis salina*: From lab-scale experiments to large-scale design. *Chem.Eng.Res.Des.*(2011), doi:10.1016/j.cherd.2011.12.002

[56] Niels-Henrik Norsker, Maria J. Barbosa, Marian H. Vermuë, René H. Wijffels. 2011. Microalgal production — A close look at the economics. *Biotechnology Advances*. 29(1): 24-27.

[57] Orlando Jorquera, Asher Kiperstok, Emerson A. Sales, Marcelo Embiruçu, Maria L. Ghirardi. 2010. Comparative energy life-cycle analyses of microalgal biomass production in open ponds and photobioreactors. *Bioresource Technology*. 101(4): 1406-1413.

[58] Pruvost, J., Van Vooren, G., Le Gouic, B., Couzinet-Mossion, A., Legrand, J. 2011. Systematic investigation of biomass and lipid productivity by microalgae in photobioreactors for biodiesel application. *Bioresource Technology*, 102(1), 150-158.

[59] E. Sevigné Itoiz, C. Fuentes-Grünewald, C.M. Gasol, E. Garcés, E. Alacid, S. Rossi, J. Rieradevall. 2012. Energy balance and environmental impact analysis of marine

microalgal biomass production for biodiesel generation in a photobioreactor pilot plant. *Biomass and Bioenergy*. 39: 324-335.

[60] H. De la Hoz Siegler, W.C. McCaffrey, R.E. Burrell, A. Ben-Zvi. 2012. Optimization of microalgal productivity using an adaptive, non-linear model based strategy. *Bioresource Technology*.104: 537-546.

[61] Jianhua Fan, Jianke Huang, Yuanguang Li, Feifei Han, Jun Wang, Xinwu Li, Weiliang Wang, Shulan Li. 2012. Sequential heterotrophy–dilution–photoinduction cultivation for efficient microalgal biomass and lipid production. *Bioresource Technology*. 112: 206-211.

[62] Evan S. Beach, Matthew J. Eckelman, Zheng Cui, Laura Brentner, Julie B. Zimmerman. 2012. Preferential technological and life cycle environmental performance of chitosan flocculation for harvesting of the green algae *Neochloris oleoabundans.BioresourceTechnology*.http://dx.doi.org/10.1016/j.biortech.2012.06.012.

[63] Dong-Geol Kim, Hyun-Joon La, Chi-Yong Ahn, Yong-Ha Park, Hee-Mock Oh. 2011. Harvest of Scenedesmus sp. with bioflocculant and reuse of culture medium for subsequent high-density cultures. *Bioresource Technology*. 102(3): 3163-3168.

[64] Hongli Zheng, Zhen Gao, Jilong Yin, Xiaohong Tang, Xiaojun Ji, He Huang. 2012. Harvesting of microalgae by flocculation with poly (γ-glutamic acid). *Bioresource Technology*. 112: 212-220.

[65] Richard M. Knuckey, Malcolm R. Brown, René Robert, Dion M.F. Frampton. 2006. Production of microalgal concentrates by flocculation and their assessment as aquaculture feeds. *Aquacultural Engineering*. 35(3): 300-313.

[66] Zechen Wu, Yi Zhu, Weiya Huang, Chengwu Zhang, Tao Li, Yuanming Zhang, Aifen Li. 2012. Evaluation of flocculation induced by pH increase for harvesting microalgae and reuse of flocculated medium. *Bioresource Technology*. 110: 496-502.

[67] Andrew K. Lee, David M. Lewis, Peter J. Ashman. 2012. Disruption of microalgal cells for the extraction of lipids for biofuels: Processes and specific energyrequirements.*Biomass.andBioenergy*.http://dx.doi.org/10.1016/j.biombioe.2012.06.034.

[68] Dang-Thuan Tran, Kuei-Ling Yeh, Ching-Lung Chen, Jo-Shu Chang. 2012. Enzymatic transesterification of microalgal oil from Chlorella vulgaris ESP-31 for biodiesel synthesis using immobilized Burkholderia lipase. *Bioresource Technology*. 108: 119-127.

[69] Jing-Qi Lai, Zhang-Li Hu, Peng-Wei Wang, Zhen Yang. 2012. Enzymatic production of microalgal biodiesel in ionic liquid [BMIm][PF6]. *Fuel*. 95:329-333.

[70] Yuchi Han, Qinxue Wen, Zhiqiang Chen, Pengfei Li. 2011. Review of Methods Used for Microalgal Lipid-Content Analysis. *Energy Procedia*. 12: 944-950.

[71] Bligh, E.G., Dyer, W.J., 1959. A rapid method of total lipid extraction and purification. *Canadian Journal of Biochemistry and Physiology* 37 (8), 911–917.

[72] Certik M , Andrasi P, Sajbidor J. Effect of extraction methods on lipid yield and fatty acid composition of lipid classes containing γ-2linolenic acid extracted from fungi. *JAOCS* ,1996 ,73 (3):357-365.

[73] Molina G E ,Robles M A ,Gimenez A , et al. Comparision between extraction of lipid and fatty acids from microalgal biomass. *JAOCS* , 1994 ,71 (9): 955-959.

[74] Tran, H.L., Hong, S.J.and Lee, C.G. 2009. Evaluation of extraction methods for recovery of fatty acids from *Botryococcus braunii* LB 572 and *Synechocystis sp.* PCC 6803. *Biotechnology and Bioprocess Engineering.* 14 (2):187–192.

[75] Richterb B E, Ezzell J L, Felix W D, et al. Comparison of accelerated solvent extraction with conventional solvent extraction for organic phosphorus pesticides and herbicides. *LC/ GC,* 1995, (13):390-398.

[76] Mou S F. Principle and application of accelerated solvent extraction. *Environmental chemistry.* 2001, 20(3):299-300.

[77] H. Evangelaras, E. Kolaiti, C. Koukouvinos. Robust parameter design: Optimization of combined array approach with orthogonal arrays.*Journal of Statistical Planning and Inference.*2006, 136(10):3698-3709.

[78] Yamini Y, Saleh A, Khajeh M. Orthogonal array design for the optimization of supercritical carbon dioxide extraction of platinum(IV) and rhenium(VII) from a solid matrix using cyanex301. *Separation and Purification Technology,* 2008, 61(1): 109-114.

[79] Georgiou S D. Orthogonal designs for computer experiments. *Journal of Statistical Planning and Inference,* 2011, 141(4): 1519-1525.

[80] Abud-Archila M., D.G., Vázquez-Mandujano MA, Ruiz-Cabrera, *et.al.* Optimization of osmotic dehydration of yam bean (Pachyrhizus erosus) using an orthogonal experimental design. *Journal of Food Engineering,* 2008,84(3): 413-419.

[81] http://www.dionex.com/en-us/products/sample-preparation/ase/instruments/lp-81379.html

[82] Schafer R B, Mueller R, Brack W, *et al.* Determination of 10 particle-associated multi class polar and semi-polar pesticides from small streams using accelerated solvent extraction. *Chemosphere,* 2008, 70(11): 1952-1960.

[83] Schafer, K. 1998. Accelerated solvent extraction of lipids for determining thefatty acid composition of biological material. *Analytica Chimica Acta.* 358 (1):69–77.

[84] Reid A M, Brougham C A, Fogarty A M, *et al.* Accelerated solvent-based extraction and enrichment of selected plasticisers and 4-nonylphenol, and extraction of tin from organotin sources in sediments, sludges and leachate soils. *Analytica Chimica Acta,* 2009,634 (2): 197-204.

[85] Zhu X L, Cai J B, Yang J, Su Q D. Determination of Organophosphate Pesticide Residues in Soil by Accelerated Solvent Extraction-Gas Chromatographyic. *Chinese Journal of Analytical Chemistry.* 2005, 33(6):821-824.

[86] Wang L, Mou Y L, Li X C. Determination of organ-phosphorus pesticide in sea sediment by accelerated solvent extraction-gas chromatography /mass spectrometry. *Chinese Journal of Health Laboratory Technology.* 2007, 17(5):769-771.

[87] Gong Y L, Sun W L, Wang S Q, Shen B. A Comparative Study on Extraction of Organic Matters in Source Rocks by Accelerated Solvent Extraction and Soxhlet Extraction. *Rock and Mineral Analysis.* 2009, 28(5):416-422.

[88] Chang C Y, Wang Y F, Ge B K, Liu C. Detection of organic chlorine pesticide residues in fruit and vegetables by using accelerated solvent extraction (ASE) method. *Port Health Control.* 2004, 96(6):25-26.

[89] DIONEX. Extraction of hydrocarbon pollutants in soil using accelerated solvent extraction (ASE) technique. *Environmental Chemistry.* 2009, 28(6):957-958.

[90] DIONEX. Extraction of PCBs in fish using accelerated solvent extraction (ASE) technique. *Environmental Chemistry.* 2008, 27(5):315-316.

[91] Wang P, Zhang Q H, Wang Y W, *et al.* Evaluation of soxhlet extraction, accelerated solvent extraction and microwave-assisted extraction for the determination of polychlorinated biphenyls and polybrominated diphenyl ethers in soil and fish samples. *Analytica Chimica Acta,* 663(1):43-48.

[92] Zhao H X, Wang L P, Qiu Y M, *et al.* Simultaneous determination of three residual barbiturates in pork using accelerated solvent extraction and gas chromatography–mass spectrometry[J]. *Journal of Chromatography* B, 2006, 840(2):139-145.

[93] Hu B Z, Song W H, Xie L P, *et al.*Determination of 33 pesticides in tea using accelerated solvent extraction/gel permeation chromatography and solid phase extraction/gas chromatography-mass spectrometry. *Chinese Journal of Chromatography,* 2008,26(1):22-28.

[94] Zheng C X, Li C E. Accelerated Solvent Extraction Technology in Traditional Chinese Medicine Active Components Analysis. *Chinese Journal of Medicinal Guide.* 2010(10):1820-1821.

[95] Zhao H Q, Chen J H, Guo X C, Zhen X L, Li X C,Wang X R. Fast extraction of alkaloids in Coptis chinensis franch by accelerated solvent extraction. *Chinese Journal of Analysis Laboratory.* 2008, 27(11):5-8.

[96] Chen J H, Li W L, Yang B J, *et al.* Determination of four major saponins in the seeds of Aesculus chinensis Bunge using accelerated solvent extraction followed by high-performance liquid chromatography and electrospray-time of flight mass spectrometry. *Analytica Chimica Acta,* 2007,596(2):273-280.

[97] Chen J H, Wang F M,Liu J, et al. Analysis of alkaloids in Coptis chinensis Franch by accelerated solvent extraction combined with ultra performance liquid chromatographic analysis with photodiode array and tandem mass spectrometry detections. *Analytica Chimica Acta,* 2008, 613(2):184-195.

[98] Song W B, Dai Y C, Xu M, Li X G, Yu J. Determination of Ginsenosides in Ginseng by ASE-SPE-LC-UV method. *Modern Scientific Instruments.* 2009,(6):104-108.

[99] Zhang Y, Wu H M, Yu J W, Zhan S L, Wang J Q. Extraction of total flavonoid from orange peel by accelerated solvent extraction. *Food Science and Technology.* 2007,(11):213-215.

[100] Pang X A, Liu W J, Sun H Z, Jin Y X, Wan Y, Ma L, Kong X Y. Application of uniform design to optimize sweet almond oil extraction by accelerated solvent extraction process. *Food and Nutrition in China.* 2007, (4):43-45.

[101] Herrero M, Pedro J, Senorans J. Optimization of accelerated solvent extraction of antioxidants from *Spirulina platensis* microalga. *Food Chemistry,* 2005, 93(3):417-423.